HANDBOOK
OF
ELECTRONIC METERS
Theory and Application

Revised and Enlarged

HANDBOOK
OF
ELECTRONIC METERS
Theory and Application

JOHN D. LENK
Consulting Technical Writer

Prentice-Hall, Inc., Englewood Cliffs, New Jersey 07632

Library of Congress Cataloging in Publication Data

Lenk, John D
  Handbook of electronic meters.

  Includes index.
  1. Electronic instruments. 2. Electronic measurements. I. Title.
TK7878.L4 1980   681'.2   80-14450
ISBN   0-13-377333-7

This revised and enlarged edition
is based on HANDBOOK OF ELECTRONIC
METERS: THEORY AND APPLICATION
© 1969 by Prentice-Hall, Inc.

Editorial/production supervision and
  interior design by Karen Skrable
Manufacturing buyer: Joyce Levatino

©1981 by Prentice-Hall, Inc., Englewood Cliffs, N.J. 07632

Printed in the United States of America

10 9 8 7 6 5 4 3 2 1

Prentice-Hall International, Inc., *London*
Prentice-Hall of Australia Pty. Limited, *Sydney*
Prentice-Hall of Canada, Ltd., *Toronto*
Prentice-Hall of India Private Limited, *New Delhi*
Prentice-Hall of Japan, Inc., *Tokyo*
Prentice-Hall of Southeast Asia Pte. Ltd., *Singapore*
Whitehall Books Limited, *Wellington, New Zealand*

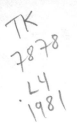
To Irene, the Sandpiper Lady and Mr. Lamb, the Magic Bunny

Contents

7   CHECKING CIRCUIT FUNCTIONS   *169*

8   SERVICING SPECIFIC CIRCUITS WITH METERS   *200*

Preface

This revised edition of the HANDBOOK OF ELECTRONIC METERS: THEORY AND APPLICATION, carries through all of the features that made the first edition so successful. That is, the revised edition bridges the gap between electronic meter theory and practical applications. The new edition thus serves the dual purpose of a basic textbook for student technicians, hobbyists, and experimenters, and a factual guidebook for experienced, working technicians. All of the chapters in the revised edition have been expanded or enlarged to include new material. Existing information has been up-dated to reflect present-day trends, especially the extensive use of digital meters. Also, much of the material from the first edition has been revised for clarification and/or simplification.

The manufacturers of meters provide instruction manuals on the operation and circuit theory of their particular instruments. Rarely, however, do these manuals give any *applications data* describing the many uses of meters. Even the training films and service courses of the largest and best-known meter manufacturers are notably lacking in such material. There are exceptions to this rule, of course.

Because the voltmeter, ohmmeter, ammeter, multimeter, and multitester are such simple instruments, and in such common use, it is generally assumed that the technician will know "automatically" the procedures for using such meters, and the capabilities and limitations. This is seldom the case. For example, how do you troubleshoot transistors "in circuit" with voltage readings? Or how do you find the resonant frequency of an LC circuit with a voltmeter, or locate the firing point of a unijunction transistor, or make a "quick-check" of a solid-state oscillator?

As in the original, this edition of the handbook fills the gap in information, and can be used to supplement the operating instructions of any meter (digital or analog), whether it be a low-cost shop type or a precision laboratory instrument. This is done by providing a variety of test, measurement, service, and troubleshooting procedures using the meter as the basic tool. These procedures are presented in "cookbook" fashion. Each procedure is preceded by a brief description of the "why" and "where" for the particular test. These descriptions offer a digest to readers who may be unfamiliar with some specialized meter applications, and want to put the step-by-step procedures to immediate use. Each operation is illustrated with test connection diagrams. Although every possible use of a meter has not been included, the practical, experience-proven applications are here.

Assuming that some readers are not familiar with the operating principles and characteristics of meters, the initial chapters give simplified presentations of these details. Chapter 1 discusses meter basics. Chapters 2 and 3 cover accessory probes, and basic operating procedures, respectively. With basics out of the way, the new edition then goes on to cover test and calibration of meters (Chapter 4), special measurement procedures (Chapter 5), using meters to check individual components (Chapter 6), and to check circuit functions (Chapter 7). Chapter 8 concentrates on servicing specific circuits with meters.

Many professionals have contributed their talent and knowledge to the revision and enlargement of the new edition. The author gratefully acknowledges that the tremendous effort to make this second edition such a comprehensive work is impossible for one person, and he wishes to thank all who have contributed directly and indirectly. The author wishes to give special thanks to the following: B&K Precision, Heathkit, Hewlett-Packard, Radio Shack, Simpson Electric Co., and Triplett Electrical Instrument Co. The author also wishes to thank Mr. Joseph A. Labok of Los Angeles Valley College for his help and encouragement.

JOHN D. LENK

1

Meter Basics

It is almost impossible to get by in any phase of electronics without some form of meter. Both hobbyists and professional technicians find it necessary to check on circuits and components—to find what voltage is available, how much current is flowing, and so on. The simplest and most common instrument that will measure the three basic electrical values (*voltage, current, and resistance*) is the *voltohmmeter* or VOM. Sometimes, the terms *multimeter* or *multitester* are used in place of VOM. There are dozens, if not hundreds, of VOMs available in all price ranges. As the price goes up, accuracy is increased, more scales or functions are added, and the scales are given greater range.

In the early days of electronics, it was common practice for the experimenter to build his own VOM. Today, because of the reduced prices and the difficulty (for practical considerations) of making accurate meter scales, the homemade VOM is almost unknown. In a way, this trend is unfortunate since much can be learned by building a VOM. As resistance is added to make a basic meter movement into a working ammeter and voltmeter, or as power and resistance are added to a basic movement for conversion to an ohmmeter, many of Ohm's and Kirchhoff's laws become practical values instead of dull theories.

The first improvement on the VOM was the *vacuum-tube voltmeter* or VTVM. Today, the VTVM has been replaced by the *transistorized* or *electronic voltmeter*. The sensitivity of these instruments is much greater than that of the VOM since electronic meters contain an amplifier. Electronic meters have another advantage over the VOM in that the electronic meter amplifier presents a high impedance to the circuit or component being measured. Thus, electronic meters draw little or no current from the circuit and have little effect on circuit operation. Those electronic meters using the *field effect transistor* or FET in their amplifiers present the highest impedance and draw the least current from the circuit, since FETs have a very high impedance compared to that of other transistors. FET meters are thus used in very sensitive electronic circuits.

The VOM, VTVM, electronic meter, and FET meter are all *analog meters*. That is, they use rectifiers, amplifiers, and other circuits to generate a current that is proportional to the quantity being measured. In turn, this current drives a meter movement. Two additional types of meters, the *differential meter* and the *digital voltmeter* (or DVM), are being used in the laboratory as a supplement to (or replacement for) the analog meters. The differential voltmeter operates by comparing an unknown voltage with a known voltage. The digital voltmeter displays measurement in discrete numerals rather than as a pointer deflection on a continuous scale as commonly used in analog instruments.

It would be almost impossible and beyond the scope of this book to describe all of the circuits used in modern meters. Many of these circuits are special-purpose. Likewise, many basic circuits are used in various combinations. Rather than attempt to describe every known meter, we shall devote the remainder of this chapter to "typical" meter circuits.

For the student, the following paragraphs also describe how a VOM is made from a basic meter movement. If it is known how an instrument operates, it will be known why an instrument can produce the desired voltage, current, or resistance information.

1-1 D'ARSONVAL MOVEMENT AND THE BASIC VOM

The simplest and most commonly used movement is the *D'Arsonval* meter movement shown in Fig. 1-1. This movement is also known as the *moving-coil galvanometer*. Early D'Arsonval movements had a core made of soft iron. A coil of very fine wire was wound on an aluminum form around the core. The iron core is now usually omitted from the movement. The coil and aluminum form function somewhat like an armature mounted on a shaft seated in jewel bearings so as to be free to turn (rotate). Springs on each end of the shaft act as current leads to the coil and help steady the coil movement.

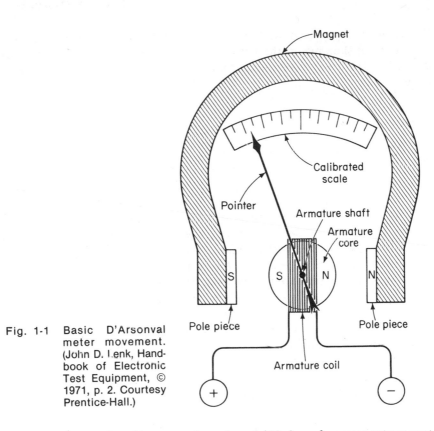

Fig. 1-1 Basic D'Arsonval meter movement. (John D. Lenk, Handbook of Electronic Test Equipment, © 1971, p. 2. Courtesy Prentice-Hall.)

The coil is placed between the poles of a U-shaped permanent magnet. One end of a pointer is fastened to the armature shaft. As the shaft rotates, the other end of the pointer moves over a calibrated dial. Current through the armature coil sets up a magnetic field that reacts with the permanent magnet's field to rotate the coil with respect to the magnet. When current passes through the coil, its magnetic field is such that the poles repel, and, since the permanent magnet cannot move, the coil rotates on its shaft. Current through the coil makes the coil turn a proportional amount. Thus, the basic meter movement is an analog device. The amount of travel of the pointer attached to the coil is related directly to the amount of current flowing through the movement. The meter scale is then related to some particular current. For example, if 1 mA is required to rotate the coil and pointer across the full scale, a half-scale reading will be equal to 0.5 mA, a quarter-scale reading will be equal to 0.25 mA, and so on.

Usually, maximum rotation of the armature (full-scale reading) is completed in less than a half turn in the clockwise direction. The complete assembly is enclosed in a glass-faced case that protects it from dust and air currents. This enclosed meter movement can be used by itself as a very sensitive ammeter. However, it is usually part of another instrument, such as a

VOM, or in a panel connected to an external circuit. In the case of a laboratory or shop meter, there is a resistor network to extend the range of the basic movement (as an ammeter) or to convert the basic movement into a voltmeter.

Basic Ammeter

The basic D'Arsonval movement, by itself, forms an *ammeter* (*ampere meter*). A true ammeter measures current in amperes. In electronics, current is more often measured in milliamperes or microamperes. Most movements used in electronic meters will produce a full-scale deflection when 1 mA (or a few microamperes in many cases) is passed through them.

A *shunt* must be connected across the meter movement if it is desired to measure currents greater than the full-scale range of the basic meter. The shunt can be a precision resistor, a bar of metal, or a simple piece of wire. Electronic-meter shunts are usually precision resistors that may be selected by means of a switch. Panel meters for heavy industrial work use metal bar shunts. Shunt resistance is only a fraction of the movement resistance. Current divides itself between the meter and shunt, with most of the current flowing through the shunt. Shunts must be precisely calibrated to match the meter movement.

Figures 1-2 and 1-3 show the two typical milliammeter range-selection circuits for VOMs. In Fig. 1-2, individual shunts are selected by the range-

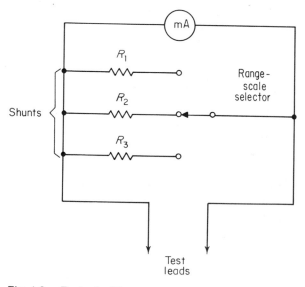

Fig. 1-2   Typical milliameter range-selection circuit (individual shunt method).

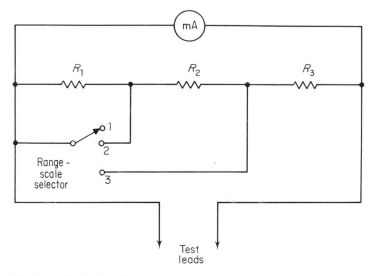

Fig. 1-3   Typical milliameter range-selection circuit (series shunt method).

scale selector. In Fig. 1-3, the shunts are cut in or out of the circuit by the selector. If the selector is in position 1, all three shunts are across the meter movement, giving the least shunting effect (most current through the movement). In position 2, resistor R_1 is shorted out of the circuit, with resistors R_2 and R_3 shunted across the movement, increasing the meter's current range. In position 3, only R_3 is shunted across the movement, and the meter reads maximum current.

It is possible to calculate the values of shunt resistance and thus convert any basic meter movement into an ammeter. The necessary equations and procedures are described in Chapter 5.

Basic Voltmeter

When the basic D'Arsonval movement is connected in *series* with resistors, a *voltmeter* is formed. The series resistance is known as a *multiplier*, since the resistance multiplies the range of the basic meter movement.

The basic voltmeter circuit is shown in Fig. 1-4. As shown, the voltage divides itself across the meter movement and the series resistance. If an 0.5-V full-scale deflection meter movement is used, and it is desired to measure a full scale of 10 V, the series resistor must drop 9.5 V. If a 100-V full scale is desired, the series resistance must drop 99.5 V, and so on. The necessary equations and procedures for calculating the values of series resistances (to convert a basic meter movement into a voltmeter) are described in Chapter 5.

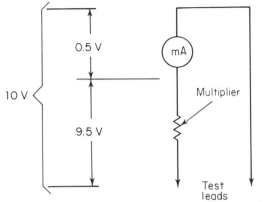

Fig. 1-4   Basic voltmeter circuit.

Figures 1-5 and 1-6 show the two typical voltmeter range-selection circuits for VOMs. In Fig. 1-5, individual multipliers are selected by the range-scale selector. In Fig. 1-6, the multipliers are cut in or out of the circuit by the selector. If the selector is in position 1, only resistor R_1 is in the circuit, giving the least voltage drop (meter will read the lowest voltage). In position 2, both R_1 and R_2 are in the circuit, giving the meter a higher voltage range. In position 3, all three resistors drop the voltage, permitting the meter to read maximum voltage.

The term *ohms per volt* is used to describe commercial VOMs. Ohms per volt is a measure of a VOM's sensitivity and represents the number of ohms required to extend the range by 1 V. For example, if the meter movement requires 1 mA for full-scale deflection, then 1000 ohms (including the movement's internal resistance) are needed for each volt that could be measured if the movement were used as a voltmeter. If the movement requires only 100 μA for full-scale deflection, then 10,000 ohms/V are needed. Thus, the more sensitive the meter movement (those requiring the least current), the higher the ohms-per-volt requirement.

Voltmeters with a high ohms-per-volt rating put less load on the circuit being measured and have a less disturbing effect on the circuit. For example, assume that a 1-V drop across a 1000-ohm circuit is to be measured with both a 100-ohms/V meter and a 20,000-ohms/V meter. A 1-V drop across a 1000-ohm circuit will produce a 1-mA current flow. With the 1000-ohms/V meter across the circuit, the 1-mA current will divide itself between the meter and the circuit. This would cut the circuit's normal current in half. With a 20,000-ohms/V meter across the same circuit, $\frac{1}{20}$ of the current will pass through the meter, and $\frac{19}{20}$ will remain in the circuit.

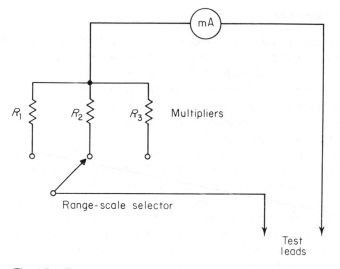

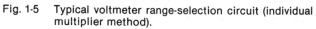

Fig. 1-5   Typical voltmeter range-selection circuit (individual multiplier method).

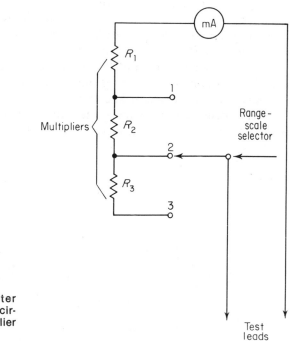

Fig. 1-6   Typical voltmeter range-selection circuit (series multiplier method).

Basic Ohmmeter

An *ohmmeter* (or resistance-measuring device) is formed when a basic meter movement is connected in series with a resistance and a power source (such as a battery in portable meters). The basic ohmmeter arrangement is shown in Fig. 1-7. Here, a 3-V battery is connected to a meter movement with a full-scale reading of 50 μA. The current-limiting resistor R has a value (60 kilohms, less meter resistance) such that exactly 50 μA flows in the circuit when the test leads are clipped together.

When there is no connection across the test leads, the current is zero. The meter's pointer rests at the "infinity" mark (∞) on the scale. When the two leads are shorted, the meter moves to the full 50-μA reading, which, on the scale, indicates a "zero ohms" reading. If a 60-kilohm resistance is connected across the leads, as shown in Fig. 1-7, the total resistance is 120 kilohms, and the meter drops to one-half of full-scale reading, or 25 μA. If the battery voltage and limiting resistor R remain constant, the pointer will always move to 25 μA whenever 60 kilohms is connected across the test leads. The 25-μA point on the meter can then be marked "60 kilohms."

With a 240-kilohm resistance across the leads, the total resistance is 300 kilohms, and the pointer drops to a 10-μA reading, since $I = E/R$, or $3/300,000 = 10$ μA. Again, if the battery voltage and the limiting resistor remain constant, the meter will always read 10 μA when a resistance of 240 kilohms is placed across the test leads. Thus, the 10-μA point on the meter scale can be marked "240 kilohms."

The ohmmeter circuit of Fig. 1-7 is then capable of measuring 60 and 240 kilohms. Any number of resistance values can be plotted on the scale, provided that resistances of known value are placed across the leads.

As discussed in Chapter 3, the ohmmeter scale of a typical VOM or multimeter is printed on the meter face along with the voltage and current

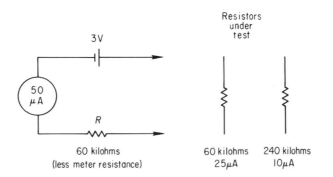

Fig. 1-7 Basic ohmmeter circuit.

8

scales. However, the ohmmeter scale is quite different from the other scales in two respects. The zero point is at the right-hand side, and the maximum resistance (usually marked "infinity" or "open") is at the left-hand side. Also, the scale is not linear (lower resistance divisions are wider, and higher resistance divisions are narrower).

The ohmmeter circuit of a typical VOM is shown in Fig. 1-8. Here, the ohmmeter has five range scales that can be selected by means of a switch. In all but the " × 1" position, a series multiplier (similar to that of a voltmeter) is connected to the circuit and drops the voltage by a corresponding amount. This reduces current flow through the entire circuit, usually by a ratio of 10 : 1, 100 : 1, 1000 : 1 or 10,000 : 1, so that the ohmmeter scale represents 10, 100, 1000, or 10,000 times the indicated amount.

No matter which scale is used, the meter and battery are in series with a variable resistor that allows the circuit to be "zeroed." As a battery ages, its output drops. Also, it is possible that with extended age or extreme temperature the resistance values or meter movement itself could change in value. Any of these conditions would make the ohmmeter scale inaccurate. The variable resistor (usually marked "zero adjust" or "zero") is included in a commercial VOM ciruit. In use, the leads are shorted together, and the variable resistor is adjusted until the meter is at zero (at the right-hand side of the ohmmeter scale). When the leads are opened, the meter then drops back to "infinity" or "open" (left-hand side), and the meter is ready to read resistance accurately.

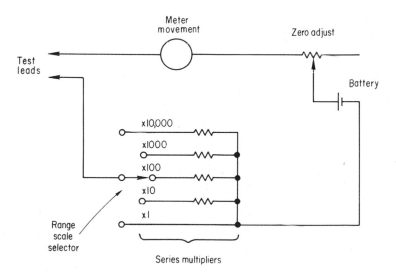

Fig. 1-8   Ohmmeter circuit of a typical VOM.

1-2 BASIC WHEATSTONE BRIDGE AND GALVANOMETER

The basic D'Arsonval meter movement can be used as a *galvanometer*. However, the term "galvanometer" has come to mean a meter where the zero of the scale is at the center, with negative current reading to the left and positive current reading to the right, as shown in Fig. 1-9. Generally, a galvanometer is used to read *proportional* positive or negative changes in a circuit rather than the actual unit value of current. The main use for such a meter is in bridge circuits, such as the *Wheatstone bridge*.

Usually, a d-c Wheatstone bridge is used to make precise resistance measurements, and a-c Wheatstone bridges are used to measure capacitance and inductance. The basic bridge circuit is shown in Fig. 1-10. Resistors R_1 and R_2 are fixed and are of known value. R_3 is a variable resistor with the necessary calibration arrangement to read the resistance value for any setting (usually a calibrated dial coupled to the variable resistance shaft). The unknown resistance value R_x is connected across terminals Y and Z, and a battery or other power source is connected across points A and C.

When switch S_1 is closed, current flows in the direction of the arrows,

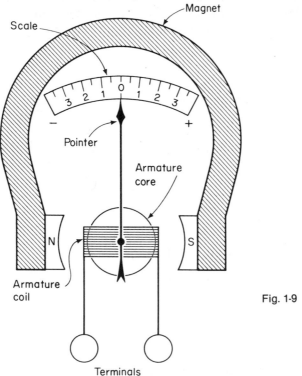

Fig. 1-9 Basic zero-center galvanometer movement. (John D. Lenk, Handbook of Electronic Test Equipment, © 1971, p. 11. Courtesy Prentice-Hall.)

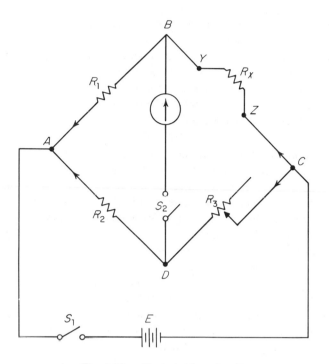

Fig. 1-10   Basic bridge circuit.

and there is a voltage drop across all four resistors. The drop across R_1 is equal to the drop across R_2. Variable resistance R_3 is adjusted so that the galvanometer reads zero (on the center scale) when switch S_2 is closed. At this adjustment R_3 is equal to R_x in resistance. By reading the resistance of R_3 (from the calibrated dial), the resistance of R_x is known.

When the variable resistance R_3 is equal to R_x, the *difference* of potential between points B and D will be zero, and no current will flow through the galvanometer. If R_3 is not equal to R_x, then B and D are not at the same voltage, and current will flow through the galvanometer, moving the pointer away from zero (on the center scale).

1-3 MEGOMETER CIRCUITS

Few, if any, commercial VOMs will provide accurate resistance measurements above 50 megohms. Shop-type VOMs will not usually provide accurate indications above 10 megohms. This is because the voltage used in the ohmmeter circuit of a typical VOM is very low. Some laboratory test meters have a built-in ohmmeter with a high-voltage power supply. The high voltage permits accurate high-resistance measurements, but such meters are usually not portable.

The megometer is essentially a portable ohmmeter with a built-in high-voltage source. The megometer, such as shown in Fig. 1-11, has two main elements: a magneto-type d-c generator to supply current for making the measurement and an ohmmeter which measures the resistance value. The generator armature is turned by a hand-crank, usually through step-up gears, to produce an output of about 500 V.

The ohmmeter portion of the megometer has two coils mounted on the same shaft but at right angles to each other. Current is fed to both coils by means of flexible connections that do not hinder rotation of the element. Coil A is the current coil, with one terminal connected to the generator negative output and the other terminal connected in series with R_1 to the test lead P_2. Test lead P_1 is connected to the generator positive output. When unknown resistance R_x is connected across P_1 and P_2, current flows from the generator through coil A, resistance R_1, and unknown resistance R_x. Resistance R_1 has enough resistance so that, even if the line terminals are short-circuited, the current coil A will not be damaged.

Coil B is the voltage coil and is connected across the generator output through resistance R. If the test leads are left open, no current will flow in coil A, and coil B alone will move the pointer. Coil B will take a position op-

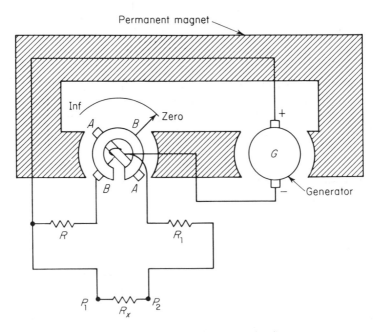

Fig. 1-11   Basic megometer circuit.

posite the gap in the core, and the pointer will indicate "infinity" or "open."

With an unknown resistance R_x connected across the test leads, current will flow in coil A. The corresponding torque developed moves the pointer away from the infinity position into a field of gradually increasing strength until the torque fields between coils A and B are equal. Variations in speed of the hand-cranked generator will not affect the megometer readings since changes in generator voltage affect both coils in the same proportion.

1-4   D'ARSONVAL ALTERNATING-CURRENT METERS

Most a-c meters are similar to d-c meters in that they are analog current-measuring devices. However, since ac reverses direction during each cycle, the basic meter movement cannot be connected directly to ac. Instead, the meter movement is connected to the a-c voltage through a rectifier. Both half-wave and full-wave rectifiers are used. However, the full-wave bridge rectifier of Fig. 1-12 is the most efficient, since a direct current will flow through the meter movement on both half cycles. The remainder of the a-c meter circuit can be identical to that of a d-c meter.

Such an arrangement will work well with alternating currents of low frequency but presents a problem as frequency increases, because the movement and multiplied resistances may load the circuit being tested. A *radio-*

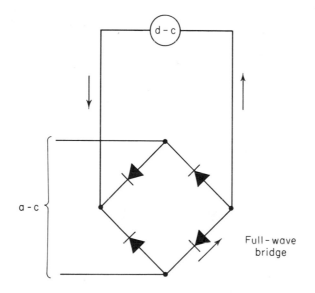

Fig. 1-12   Basic a-c meter circuit with full-wave bridge.

frequency probe can be connected ahead of the basic meter circuit to over-come this condition. In most VOMs, the RF probe (Fig. 1-13) is a slender metallic prod on the end of an insulated rod or handle connected to the in-strument terminal through a flexible insulated lead. In operation, capacitor *C* is used to protect the diode from damage by any dc in the circuit under test (by blocking dc but passing ac). The diode rectifies the RF voltage and develops a d-c output voltage across the load resistor *R*. This voltage is then measured in the normal manner. The subject of meter probes is discussed further in Chapter 2.

A-c meter scales also present a problem. As shown in Fig. 1-14, there are four ways to measure an a-c voltage. We can measure the average, RMS or effective, peak, or peak-to-peak voltage.

Peak voltage is measured from the crest of one half cycle, while *peak-to-peak voltage* is measured from the crests of both half cycles. However, the direct current to the meter movement will be less than the peak alter-nating current, since the voltage and current drop to zero on each half cycle.

With a full-wave bridge rectifier, the current or voltage will be 0.637 of the peak value (a half-wave rectifier will deliver 0.318 of the peak value). This is known as *average* value, and some meters are so calibrated. Most meters have RMS, or root-mean-square, scales. In an RMS meter, the scale indicates 0.707 of the peak value (assuming the usual full-wave rectifier is used). This value is the *effective* value of an alternating current.

A direct current flowing through a resistor produces heat. So does an alternating current. The effective value of an alternating current or voltage is that value which will produce the same amount of heat in a resistor as direct current or voltage of the same numerical value. The term "RMS" is used since it represents the square root of the average of the squares of all instantaneous values in a *perfect sine wave*. Since nearly pure sine waves are frequently measured, this mathematical representation is of particular im-portance, and it is important to know that the effective value of a sine wave (its heat-producing equivalent of dc) is 0.707 of the peak value. The subject of meter scales is discussed further in Chapter 3.

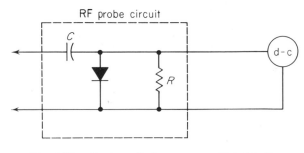

Fig. 1-13   Basic radio-frequency probe circuit.

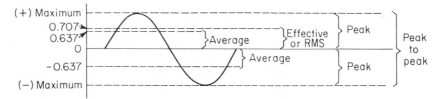

Given	Average	Effective (RMS)	Peak	Peak-to-peak
Average	——	1.11	1.57	1.271
Effective (RMS)	0.900	——	1.411	2.831
Peak	0.637	0.707	——	2.00
Peak-to-peak	0.3181	0.3541	0.500	——

Fig. 1-14   Relationship of average, effective or RMS, peak, and peak-to-peak values for alternating-current sine waves.

1.5 CLIP-ON METERS

One particular meter that is unique to a-c measurements is the clip-on meter shown in Fig. 1-15. Alternating currents set up alternating fields around a wire or conductor. These currents can be picked up by a coil of wire around the conductor, stepped up through a transformer, and measured by a volt-meter. A clip-on meter, complete with built-in coil, transformer, and meter movement, is particularly useful where conductors are carrying heavy currents and where it is not convenient to open the circuit to insert an ammeter.

Current probes are similar to clip-on meters, except that probes are generally used in conjunction with an amplifier to measure small currents. Most electronic laboratories use current probes rather than clip-on meters. A typical probe clips around the wire carrying the current to be measured and, in effect, makes the wire the one-turn primary of a "transformer" formed by ferrite cores and a many-turn secondary within the probe. The signal induced in the secondary is amplified and can be applied to any suitable a-c voltmeter for measurement. In commercial units, the amplifier constants are chosen so that 1 mA in the wire being measured produces 1

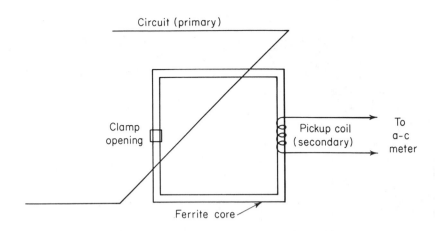

Circuit (primary)

Clamp
opening

Pickup coil
(secondary)

To
a-c
meter

Ferrite core

Fig. 1-15   Basic clip-on meter or current probe circuit.

mV at the amplifier output. In this way, current can be read directly on the voltmeter.

1-6 THERMOCOUPLE METERS

The thermocouple meter measures dc, ac, and even RF. Figure 1-16 shows the basic circuit used in thermocouple meters. When bars of two dissimilar metals are connected at one end and heat is applied to the connected ends, a d-c voltage is developed across the open ends of the two dissimilar metals. This voltage is directly proportional to the temperature of the wires in the heated junction. The generation of d-c voltage by heating the junction is called *thermoelectric* action, and the device is called a *thermocouple*.

Any two bars of dissimilar metals will produce a voltage across their open ends when their junction is heated. However, certain combinations of metals will produce various specific voltages for each degree of temperature difference. Commonly used metal combinations are: copper-constantan, iron-constantan, chromel-constantan, chromel-alumel, and platinum-rhodium. Tables are available that show the voltages produced by each of the various metal combinations at specific temperatures.

An electric current passing through a wire or conductor will produce heat in that wire in *proportion to the square* of the current. Therefore, if a current is passed through the junction of a thermocouple, heat will be generated in the wires, and a voltage will be developed at the open ends. If a calibrated meter movement is connected to the free ends of the thermocouple, the generated voltage can be measured. Of course, the meter scale must be calibrated to relate the reading to the amount of current passing through the thermocouple and heating the junction, rather than the voltage produced by the thermocouple.

16

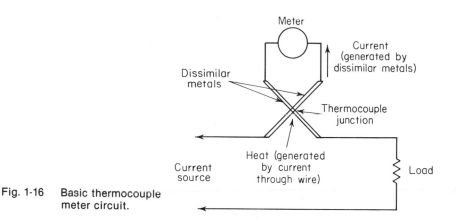

Fig. 1-16   Basic thermocouple
meter circuit.

The direction of current in the thermocouple has no effect on the heating of the junction, so the instrument can measure dc, ac, or RF. When measuring very low currents or very high frequencies (some thermocouple systems operate up to 50 MHz), the thermocouple junction is usually sealed in a vacuum similar to the filament of a vacuum tube. This gives the greatest amount of heat for a minimum amount of current.

1-7 HOT-WIRE AMMETER

The hot-wire ammeter is one of the older methods used to measure small a-c or RF currents. As shown in Fig. 1-17, the alternating current (or RF) travels through a fine wire stretched horizontally between points A and B. Another wire is attached to point C on the horizontal wire and is fastened at point D. A fine thread attached to point E of this second wire is also attached to the indicator at point F and is tied to a small spring at point G. As the current passes through the wire AB, the resistance causes the wire to heat and expand. This slight expansion lengthens the wire, and the spring G then pulls the pointer to a corresponding value on the scale. The heating effect is proportional to the square of the current through the wire. Therefore, the calibrations on the scale of a hot-wire ammeter are not equally spaced, rather the distance between the calibrations increases as the square increases.

1-8 IRON-VANE METER

The iron-vane meter will also measure ac, but is not too effective for RF. Therefore, the iron-vane meter is usually limited to heavy industrial work. As shown in Fig. 1-18, the iron-vane meter has two soft-iron magnetized pieces or vanes mounted inside a coil. One of the vanes is fixed, and the other is free to move. A shaft and pointer is attached to the moving vane.

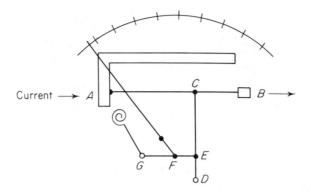

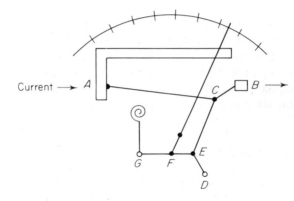

Fig. 1-17   Basic hot-wire ammeter.

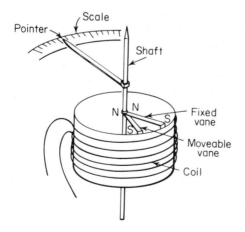

Fig. 1-18   Basic iron-vane meter.

As current flows through the coil, the two vanes become magnetized Since the vanes are magnetized with like poles at the same ends, the vanes tend to repel each other. The free vane moves away from the fixed vane. This turns the shaft and moves the indicator across the calibrated dial.

Even though the direction of alternating current changes on each half cycle, the two vanes are always magnetized alike and will repel each other. Reversals in current have no effect on the pointer indication. Since the strength of the magnetic field in the two vanes is directly dependent on the amount of current passing through the coil, the value of the current is indicated by the pointer on the uniformly calibrated dial.

1-9 DYNAMOMETER

The dynamometer is another meter movement that will measure either ac or dc and can be used either as a voltmeter or as an ammeter. Figure 1-19 shows the ammeter connection, and Fig. 1-20 shows how a dynamometer can be used as a voltmeter.

As shown in Fig. 1-19, the polarity of all the coils is such that the moving coil will rotate to the right on the first half cycle of alternating current. The distance moved is determined by the coil spring attached to the pointer

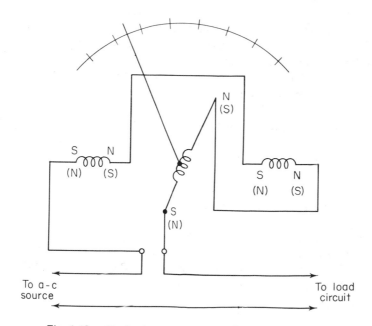

Fig. 1-19   Basic dynamometer used as an ammeter.

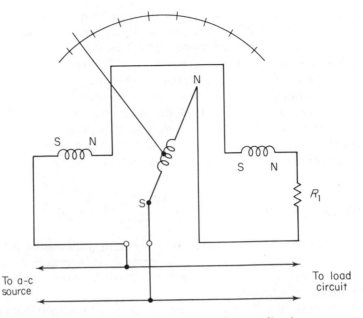

Fig. 1-20   Basic dynamometer used as a voltmeter.

shaft. When the tension on the spring becomes equal to the pull of the magnetic fields, the pointer will come to the rest.

On opposite half cycles of the alternating current, the polarity of *all coils* will reverse. When this occurs, the same amount of force is still exerted to turn the movable coil, and the direction of rotation is the same as before (to the right, or clockwise). The same will occur when direct current is applied to the coils. With either ac or dc, the readings are *approximately equal to the square* of the current.

The dynamometer is somewhat limited as to current capacity. However, the dynamometer can be used quite effectively as a voltmeter when a series resistor is added to the circuit as shown in Fig. 1-20.

1-10 LABORATORY ANALOG METERS

Laboratory analog meters operate by producing a current proportional to the quantity being measured, as do shop-type VOMs. However, laboratory meters include many circuit refinements to improve their accuracy and stability. Also, there are special-purpose analog meter circuits which are unique to laboratory instruments.

Basic Electronic Meters

The basic electronic meter circuit shown in Fig. 1-21. The amplifier can be either vacuum-tube (for VTVM) or transistor. A typical VTVM amplifier provides about 11 megohms input impedance, whereas a FET meter amplifier can provide up to about 100 megohms. A two-junction transistor provides an input impedance between these limits. The amplifier is generally direct-coupled. That is, there are no coupling capacitors between the amplifier stages.

When the basic circuit is used as a voltmeter, resistance R has a large value (usually several megohms) and is connected in parallel across the voltage circuit being measured. Because of the high resistance, very little current is drawn from the circuit, and operation of the circuit is not disturbed. The voltage across resistance R raises the voltage at the amplifier input from the zero level. This causes the meter at the amplifier output to indicate a corresponding voltage.

When the basic circuit is used as an ammeter, resistance R has a small value (a few ohms or less) and is connected in series with the circuit being measured. Because of the low resistance, there is little change in the total circuit current, and operation of the circuit is not disturbed. Current flow through resistance R causes a voltage to be developed across R, which raises the amplifier input level from zero and causes the meter to indicate a corresponding (current) reading.

One of the most common circuits used in electronic meters is shown in Fig. 1-22. This is essentially a differential amplifier, in which the voltage to be measured is applied to one input, and the other input is grounded. The zero-set resistance is adjusted so that the meter reads zero when no input voltage is applied. When the voltage to be measured is applied across the input resistance, the circuit is unbalanced, and the meter indicates the proportional unbalance as a corresponding voltage reading. One of the reasons for

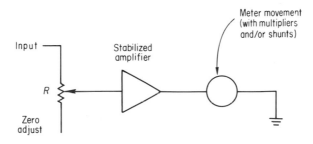

Fig. 1-21   Basic electronic voltmeter circuit.

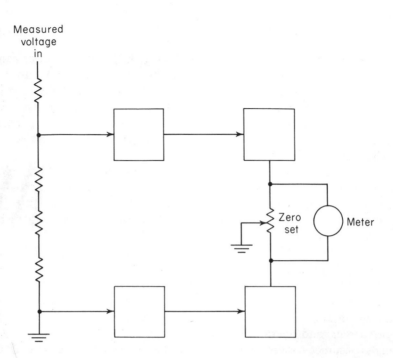

Fig. 1-22   Basic voltmeter with differential amplifier.

using the differential amplifier circuit is to minimize drift due to power supply changes. (The drift problem is discussed further in a later paragraph.)

The amplifier in Figs. 1-21 and 1-22 performs three basic functions.
First, the effective sensitivity of the meter movement is increased. An amplifier changes the measured quantity into a current of sufficient amplitude to
deflect the meter movement. Thus, a few microvolts that would not show
up on a typical VOM could be amplified to several volts to deflect any meter
movement.

The second function of the amplifier is to increase the *input impedance*
of the meter so that the instrument draws little current from the circuit
under test.

The third amplifier function is to limit the maximum current applied
to the meter movement. Therefore, there is little danger when unexpected
overloads occur which could burn out the meter movement.

Drift Problems in Electronic Meters

One of the problems common to any direct-coupled amplifier is drift due to
power-supply change. The amplifier cannot tell the difference between
power-supply change and change in the voltage being measured. This is

especially aggravated when an electronic meter is used to measure small voltages.

A common technique for eliminating drift in electronic meter circuits is to convert the d-c voltage being measured into an a-c voltage (Fig. 1-23). This is done by alternately applying and removing the d-c input voltage through a "chopper," amplifying the "chopped" signal in an a-c amplifier, and then synchronously rectifying the amplifier output back to a d-c voltage for application on a meter movement. Overall d-c feedback ensures accuracy of the d-c gain. Therefore, d-c drift is limited to a value set by the input chopper. An electromechanical chopper can be used, but some form of electronic chopper (such as the photoconductive chopper shown in Fig. 1-23) is used more often. In a typical photoconductive chopper, the d-c input voltage is converted to a comparable a-c voltage by illuminating a group of photoconductive resistors periodically. This results in a low-noise, high-impedance chopper action.

In the circuit of Fig. 1-23, photoconductive resistors are illuminated by a flasher, such as a neon lamp driven by a 60-Hz line source. The input and output signals are synchronized since they are both driven by the same

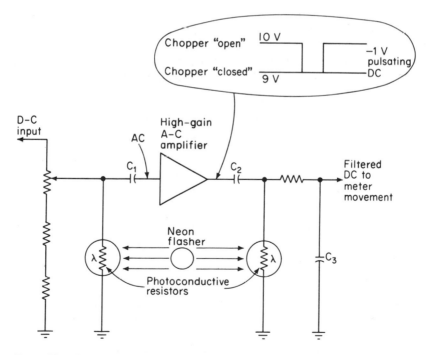

Fig. 1-23   Basic circuit for eliminating drift in electronic meters. (John D. Lenk, Handbook of Electronic Test Equipment, © 1971, p. 17. Courtesy Prentice-Hall.)

source. When the photoconductive resistors are illuminated, they provide low resistance and act as a short circuit. Without light, the resistors provide very high resistance and appear as an open circuit.

With light, the input to the amplifier is grounded (zero input). One side of capacitor C_2 is also grounded. The opposite side of C_2 is connected to the amplifier output circuit. Capacitor C_2 charges up to the output circuit value. When the flasher light is removed on alternate half cycles of the 60-Hz driving source, the d-c voltage to be measured is applied to the amplifier input. This causes the output to drop by an amount proportional to the input signal. Capacitor C_2 then discharges into the meter circuit by a corresponding amount. For example, assume that C_2 is charged to 10 V with no input and that the amplifier output drops to 9 V when the input is applied (chopper open). Under these conditions, the output is then (ideally) a square wave equal to -1 V.

Resistance Measurements in Electronic Meters

In a VOM, resistance is measured by applying a known voltage to the unknown resistance and then measuring the current passing through the circuit. When voltage and current are known, resistance can be computed. In actual practice, computation is unnecessary since the resistance scale of the meter is precalculated.

Most electronic meters use a modified procedure for resistance measurement. As shown in Fig. 1-24, the current in the circuit depends on the

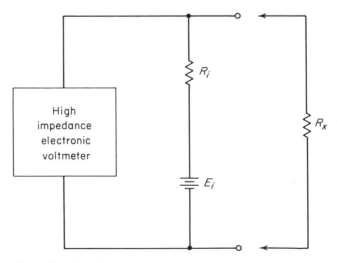

Fig. 1-24   Basic resistance measurement procedure in electronic meters.

series combination of the unknown resistor R_x and the internal resistor R. This means that both the voltage and current in the external circuit will change according to the value of the unknown. The resistance scales of the meter are calibrated for the measurement of this unknown resistance.

If R_x is infinite, the meter reads the full battery voltage. Full-scale deflection corresponds to a resistance of infinity. If R_x is zero (short circuit), the meter reads zero. The midscale range occurs when R_x equals R_i.

The resistance R_i, included as part of the ohmmeter circuit, provides a convenient means of changing the range of the instrument. When values of low resistance are being measured, the resistance of the ohmmeter leads (included in the total resistance measurement) can add considerable error. To overcome this condition, the circuit can be altered to that shown in Fig. 1-25. Here, the resistance of the current-carrying leads is calibrated as part of R_i, and the resistance in the voltmeter leads is small compared with the high input impedance of the metering circuit.

An *external power source* is often used in laboratory work when it is necessary to measure very high or very low resistances, as discussed in chapter 5.

For high resistances, a high voltage is applied to the unknown, and the current is measured on a sensitive current meter. High-resistance measurements can be disturbed by the impedance of the measuring voltmeter when this impedance is comparable to the resistance being measured. Many laboratory meters account for this by adjusting the value of R_i on the high-resistance ranges to compensate for the voltmeter input impedance. For example, on a 100-megohm scale the value of R_i is actually 200 megohms. The

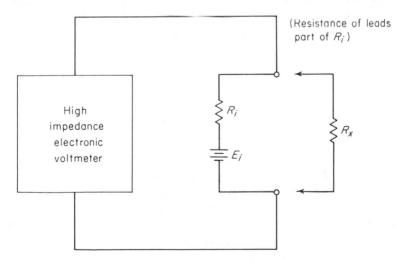

Fig. 1-25   Improved resistance measurement procedure for electronic voltmeters.

parallel combination of the 200-megohm R_i and the 200-megohm input impedance of the meter gives an effective internal impedance of 100 megohms.

To measure very low resistances, such as those found in short lengths of wire or in relay contacts, a constant-current source may be used to supply a fixed amount of current through the unknown resistance. The voltage drop across the resistance being measured is then indicated by a sensitive voltmeter. Resistance measurements as low as 1 microohm can be made by this technique.

Laboratory Meter Scales and Movements

In high-accuracy meters, a taut-band suspension is substituted in place of the pivots and jewels of the shop-type meter. The moving coil in the taut-band meter mechanism is suspended on a platinum-alloy ribbon, eliminating friction and problems concerning repeated measurements.

Laboratory meter faces are custom-calibrated and photographically printed to match exactly the linearity characteristics of each individual meter movement at all points. This eliminates the tracking error found on many mass-produced meters.

By combining custom-calibrated scales with taut-band suspension, the possibility of mechanical error is kept to an absolute minimum.

A-C Measurements in Electronic Meters

Electronic meters for measuring a-c voltages also use an amplifier with the meter movement but add a rectifier circuit to convert the ac into dc. Most shop-type meters are RMS-*reading* instruments. This is also true of laboratory meters, although it is possible to use average-, peak-, or peak-to-peak-reading meters for special applications.

Although a meter may be RMS-*reading*, it is usually average- or peak-*responding*. That is, the scale reads RMS values, but the meter movement operates on an average or peak value.

Figure 1-26 shows the basic circuit of an average-responding meter. Here, the a-c signal is amplified in a gain-stabilized a-c amplifier and then is rectified by the diodes. The resulting current pulses drive the meter. The meter deflection is proportional to the average value of the waveform being measured (even though the scale may or may not read RMS).

The peak-responding voltmeter shown in Fig. 1-27 places the rectifier in the *input circuit* where it charges the small input capacitor to the peak value of the input signal. This voltage is passed to a d-c amplifier which drives the meter. Meter deflection is proportional to the peak amplitude of the input waveform. The meter scale can be calibrated in RMS or peak voltage as required.

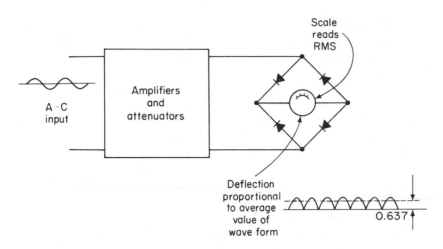

Fig. 1-26   Basic average-responding (RMS-reading) a-c voltmeter. (Courtesy Hewlett-Packard.)

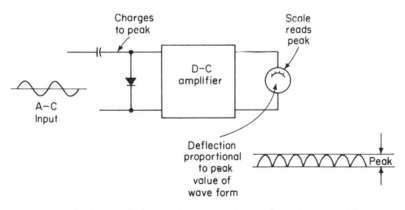

Fig. 1-27   Basic peak-responding a-c electronic voltmeter. (Courtesy Hewlett-Packard.)

Both of these meters (average-responding and peak-responding) are used primarily for sine-wave measurements. Consequently, both meters may be in error if the measured signal is not a pure sine wave. The phase and amplitude of the harmonics present in a nonsine wave will affect both peak and average values of the waveform. This can upset the RMS calibrations. The peak-responding meter is affected more by distortion than the average-responding circuit. However, the peak-responding meters are used for higher-frequency measurements since their input capacitance is lower than that of the average-responding meters. (The capacitance to ground of the input circuit and probe of a voltmeter must be included as part of the input impedance. This capacitance acts as a high-frequency bypass to the input resistance and limits the frequency range of the a-c voltmeter.)

If highly complex waveforms are to be measured, a true RMS-*responding* voltmeter should be used. Such a circuit is shown in Fig. 1-28. Here, the complex waveform is used to heat the junction of thermocouples.

One of the major problems in this technique is the nonlinear behavior, as well as slow response and possible burnout, of the thermocouple. Nonlinear behavior complicates calibration of the indicating meter. This difficulty can be overcome by the use of two thermocouples mounted in the *same thermal environment*. Nonlinear effects in the measuring thermocouple are offset by similar nonlinear operations of the second thermocouple.

As shown in the circuit of Fig. 1-28, developed by Hewlett-Packard, the amplified input signal is applied to the measuring thermocouple, and a d-c feedback voltage is fed to the balancing thermocouple. The d-c voltage is obtained from the voltage output *difference* between the thermocouples. The circuit can be considered as a feedback-control system which matches the heating power of the d-c feedback voltage to the input waveform heating power. Meter deflection is proportional to the d-c feedback voltage, which, in turn, is proportional to the RMS of the input voltage (no matter what the waveform). Therefore, the meter indication is linear.

It is possible to measure peak-to-peak voltages with an electronic voltmeter. The circuit is similar to that of the peak-responding circuit of Fig. 1-27. However, the input capacitor is charged to the peak-to-peak value by a full-wave rectifier. Also, the scale is calibrated to read directly in peak-to-peak values.

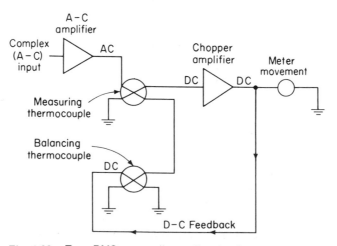

Fig. 1-28 True RMS-responding voltmeter for measurement of complex waves. (Courtesy Hewlett-Packard.)

Power Measurements

There are several methods used to measure power in electrical circuits. Although the methods are relatively simple, it is necessary to know the characteristics of power in electrical circuits to understand how the measurements are made.

The basic method for measuring the electrical power consumed by a load is to connect an ammeter in series with this load, and a voltmeter across the load. Then, since the power (in *watts*) is equal to the product of the current flowing through the load (in *amperes*) and the voltage (in *volts*) applied across the load, the power can be determined by multiplying the readings on the two meters.

The power, in watts, is thus indicated in d-c circuits (where there is no *phase difference* between current and voltage), or in a-c circuits where only a resistive load is concerned, and the current and voltage are in phase. In an a-c circuit, where the load has an *inductive* or *capacitive* component, a new factor, called *reactance*, is introduced. As a result, the current and voltage of the circuit no longer are in phase, and the phase difference is a function of the amount of inductance or capacitance (or both).

There are two types of power in an a-c circuit with and inductive or capacitive component. One is the *true* or useful power, which can do useful work (such as causing a motor to rotate). The other power is useless or *reactive* power, which can do no useful work. Total power, which is known as *apparent power*, is the sum of the true power and the reactive power. The ratio between the true power and the apparent power in an a-c circuit is known as the *power factor* of the circuit.

The power factor equals the true power divided by the apparent power, and the true power equals the apparent power multiplied by the power factor. This power factor, which is always less than 100 percent, is a function of the phase difference between the current and voltage in the circuit. The apparent power in the circuit may be measured by the basic voltmeter/ammeter method. However, to distinguish apparent power from true power, apparent power is stated in units of volt-amperes. True power, which is stated in units of watts, is measured by a *wattmeter*.

Figure 1-29 shows the circuit of a wattmeter used to measure power in 60-Hz power line circuits. As shown, the instrument consists of fixed and moving coils. The two fixed coils, with few turns of heavy wire, are connected in series with the line, and their magnetic fields are a function of the current in the circuit. The moving coil, with many turns of fine wire, together with a series multipler resistor R, is connected in parallel with the load, and the magnetic field is a function of the voltage across the load.

Rotation of the moving coil, and the pointer attached to the coil, is a function of the product of the instantaneous values of current and voltage

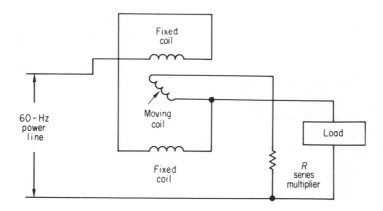

Fig. 1-29   How a single-phase wattmeter is connected to measure the true power consumed by a load.

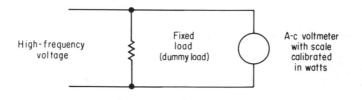

Fig. 1-30   Wattmeter used to measure power in higher frequency systems.

in the circuit. The scale is calibrated to indicate this product in watts. Similar circuits are available to measure power in control systems using three-phase power. A separate set of fixed and moving coils is used for each phase. The scale pointer is connected to all moving coils, and movement of the power is determined by the power in all three phases. Thus, the scale over which the pointer moves is calibrated to indicate the total true power of the entire system.

Figure 1-30 shows the circuit of a wattmeter used to measure power in higher-frequency systems (such as the output of a radio transmitter). Here, the output of the circuit is applied to a fixed load (often called a *dummy load*). Typically, the load is a noninductive (not wire-wound) carbon or composition resistor. The value of this load resistor is approximately equal to the output impedance of the circuit being measured. The meter, which is actually an a-c voltmeter (possibly a thermocouple type), has a scale calibrated in watts.

30

1-11 DIFFERENTIAL METERS

The basic concept of differential voltage measurement is to apply an unknown voltage against one that is accurately known and to measure the *difference between the two* on an indicating device. If the known voltage is adjusted to the exact potential of the unknown voltage, one can determine the unknown quantity being measured as accurately as the known voltage (or reference standard).

Measurements made by the differential voltmeter technique (sometimes called a potentiometric or manual voltmeter measurements) are recognized as one of the most accurate means of relating an unknown voltage to a known reference. In practice, these measurements are made by adjusting a *precision resistive divider* to divide down an accurately known reference voltage. The divider is adjusted to the point where the divider output equals the unknown voltage as indicated by the null voltmeter of Fig. 1-31.

The unknown voltage is determined to an accuracy limited only by the accuracies of the reference voltage and the resistive divider. Meter accuracy is of little consequence, since the meter serves only to indicate any residual difference between the known and unknown voltages.

A high-voltage standard is required to measure high voltage. This need may be overcome by inserting a voltage divider between the source and the null meter, as shown in Fig. 1-32. This, however, results in relatively low

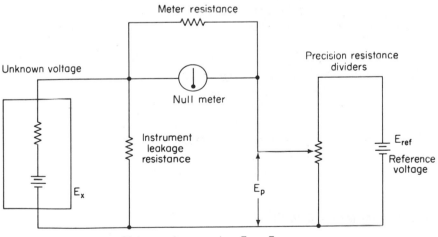

Meter reads zero when $E_p = E_x$

Fig. 1-31   Basic differential voltage measurement. (Courtesy Hewlett-Packard.)

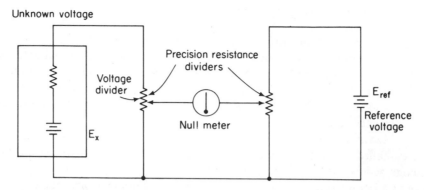

Unknown voltage

Precision resistance dividers

Voltage divider

Null meter

E_x

E_{ref}

Reference voltage

Fig. 1-32   Potentiometric method of measuring unknown voltages. (Courtesy Hewlett-Packard.)

input resistances for voltages higher than the reference standard. This low input resistance is undesirable because accurate measurement may not be obtained if substantial current is drawn from the source being measured. Many differential meters offer input resistance approaching infinity only at a null condition, and then only if an input voltage divider is not used.

To overcome these conditions, Hewlett-Packard has developed a differential voltmeter with an *input isolation stage*. A simplified diagram of the circuit is shown in Fig. 1-33. Isolation is accomplished by means of an *operational amplifier*, or *op-amp*, between the measurement source and the measurement circuits. IC (or integrated circuit) op-amps are often used for this purpose. The amplifier ensures that the high input impedance is maintained regardless of whether the instrument is adjusted for a null reading or not.

A further advantage provided by the amplifier is that the resistive voltage divider (which permits voltages as high as 1000 V to be compared to a precision 1-V reference) may be placed at the output of the amplifier, rather than being in series with the measured voltage source. This isolation permits the instrument to have high impedance on all ranges.

As shown in Fig. 1-33, the instrument also serves as a precision d-c source. The need for precision d-c sources with differential meter measurements of resistances is discussed later in this section.

Differential Meter Circuits

Differential meter circuits are used for measurement of both a-c and d-c voltages, as well as voltage ratios. When making d-c voltage measurements, there are cases where the absolute value of the voltage is of little interest. Instead, the point of interest is the value in relationship to some other voltage

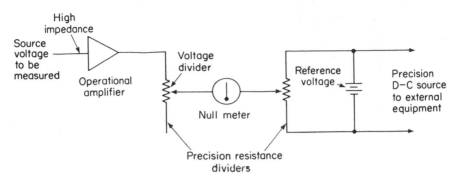

Fig. 1-33 Potentiometric voltmeter with high-impedance input amplifier. (Courtesy Hewlett-Packard.)

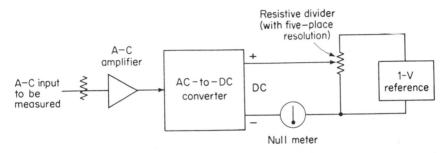

Fig. 1-34 Basic differential voltmeter in a-c mode of operation. (Courtesy Hewlett-Packard.)

level, or the ratio to some other level. This ratio, expressed as N = voltage A/voltage B = ratio, appears often in engineering work.

Figure 1-34 is a block diagram of a typical differential voltmeter *in the a-c voltage mode of operation.* This circuit uses a precision rectifier to convert the unknown a-c voltage directly to d-c voltage (equivalent to the average value of the a-c voltage), and the resulting d-c voltage is read to five-place resolution by the potentiometric voltmeter technique. The measurement is straightforward in that the a-c voltage remains connected to the converter at all times and can be monitored continuously.

Figure 1-35 is a block diagram of a typical differential voltmeter *in the d-c voltage mode of operation.* To accurately measure a d-c voltage using the differential principle, the unknown voltage is connected across the series combination of the electronic voltmeter and the 500-V reference supply. The reference voltage is then adjusted with five voltage readout dials (mechanically coupled to a Kelvin-Varley voltage divider) until the reference voltage matches the unknown voltage, as indicated by a null on the electronic voltmeter. The unknown voltage is then read from the five reference voltage-readout dials.

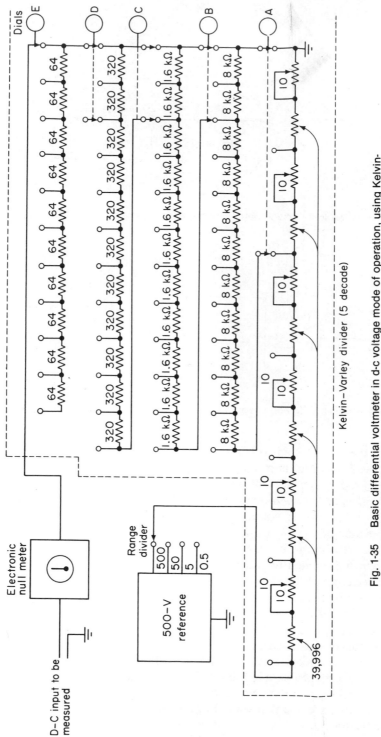

Fig. 1-35   Basic differential voltmeter in d-c voltage mode of operation, using Kelvin-Varley divider. (John D. Lenk, Handbook of Electronic Test Equipment, © 1971, p. 68. Courtesy Prentice-Hall.)

1-12 DIGITAL METERS

To fully understand operation of digital meters it is necessary to have a full understanding of digital logic circuits. These include gates, amplifiers, switching elements, delay elements, binary counting systems, truth tables, registers, encoders, decoders, D/A converters, A/D converters, adders, scalers, counter/readouts, and so on. Full descriptions of these devices are contained in the author's *Logic Designers Manual* (Reston, Va.: Reston Publishing Company, 1977). However, it is possible to have an adequate understanding of digital meters if the reader can follow simplified block diagrams, as presented here.

In actual practice, most digital meters are made up of "logic building blocks" (gates, registers, counters, etc., often in IC form) that are interconnected to perform various functions (mathematical operations, conversion, readout, etc.). If the reader understands operation of the individual building blocks, he or she can then understand operation of the complete instrument. Thus, operating principles of complex digital meters can be presented in block form or in logic-diagram form, with blocks and logic symbols representing the building blocks. It should be noted that this method of presentation is followed by most manufacturers of digital meters in their instruction manuals.

It should be further noted that a knowledge of electronic *counter/ readouts* is necessary to fully understand digital meters. This is because a digital meter performs two functions: (1) conversion of voltage (or other quantity being measured) to time or frequency (usually in the form of pulses), and (2) conversion of the time or frequency data to a digital readout. In effect, a digital meter is a conversion circuit (voltage to time, etc.) plus an electronic counter/readout. For this reason, counter/readouts are discussed more fully at the end of this section.

Basic Digital Meter

The most popular digital meter is the DMM or digital multimeter (also known as the DVM or digital voltmeter, although most DVMs are capable of measuring current and resistance as well). Such instruments display measurements as *discrete numerals*, rather than as a pointer deflection on a continuous scale. Direct numerical readout reduces human error, eliminates parallax error (discussed further in Chapter 3), and increases reading speed. Automatic polarity and range-changing features on some digital meters reduce operator training, measurement error, and possible instrument damage through overload.

Figure 1-36 shows the operating controls and readout of a typical digital multimeter. Note the simplicity of controls. Once the power is turned

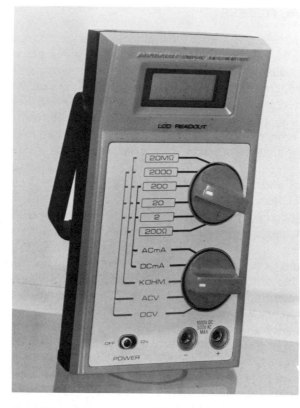

Fig. 1-36   Radio Shack LCD Digital Multimeter. (Courtesy Radio Shack.)

on, the operator has only to select the desired function and range before connecting the meter to the circuit. The readout is automatic. On some digital multimeters, the range is changed automatically, further simplifying operation.

Ramp-Type Digital Meter

The block diagram of a typical ramp-type digital voltmeter is shown in Fig. 1-37. Conversion of a voltage to a time interval is illustrated by the timing diagram of Fig. 1-38. The operating principle of the ramp-type digital voltmeter is to measure the time required for a linear voltage ramp to change from a value equal to the voltage being measured to zero (or vice versa). The *time period* is measured with an electronic counter and is displayed on a decade readout. The ramp-type meter is essentially a voltage-to-time converter, plus a counter/readout.

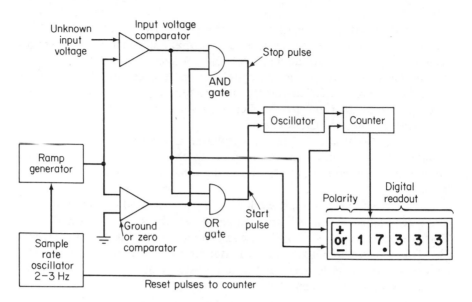

Fig. 1-37   Basic ramp-type digital voltmeter circuit. (Courtesy Hewlett-Packard.)

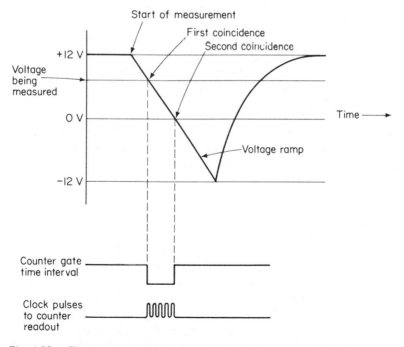

Fig. 1-38   Timing diagram showing voltage-to-time conversion. (Courtesy
Hewlett-Packard.)

At the start of a measurement cycle (there are usually two or three measurement cycles per second), a ramp voltage is generated. The ramp voltage is compared continuously with the voltage being measured. At the instant the two voltages become equal, a coincidence circuit generates a pulse that opens a gate. The ramp continues until a second comparator circuit senses that the ramp has reached zero volts. The output pulse of this comparator closes the gate.

The time duration of the gate opening is proportional to the input voltage. The gate allows pulses to pass to the counter circuit, and the number of pulses counted during the gating interval is thus a measure of the voltage.

As shown in Fig. 1-37, the voltage ramp is generated by a ramp or sawtooth generator, which, in turn, is triggered by the sample rate oscillator. This oscillator serves as a time base and produces pulses at the rate of 2 or 3 Hz. These pulses are also sent to the counter circuits to clear any readings (return the readings to zero) at the same rate. This makes it possible to monitor a steadily changing voltage.

The ramp voltage is compared with both the input or unknown voltage (in the input comparator) and zero voltage (in the zero or ground comparator). The output of the two comparators is connected to an AND gate (for stop pulses) and an OR gate (for start pulses). The AND and OR gates control operation of the counter circuit oscillator. Coincidence of the ramp voltage with either the voltage being measured (or with zero volts) starts the oscillator. (The AND gate is disabled, and the OR gate is enabled.) With the oscillator on, the electronic counter starts the count.

The elapsed time, as indicated by the count on the readout, is proportional to the time the ramp takes to travel between the unknown voltage and 0 V (or vice versa in the case of a negative input voltage to be measured). Therefore, the count is equal to the input voltage. The order in which the pulses come from the two comparators indicates the polarity of the unknown voltage. This triggers a readout that indicates plus, or minus, as required.

Staircase Ramp Digital Meter

The staircase-ramp type of digital meter, shown in Fig. 1-39, is an improvement over the basic ramp type. This digital meter makes voltage measurements by comparing the input voltage to an internally generated *staircase ramp* voltage, rather than to a linear ramp. When the input and the staircase ramp voltages are equal, the comparator generates a signal to stop the ramp and the count. The instrument then displays the number of counts that were necessary to make the staircase ramp equal to the input voltage. At the end of the sample (the sample rate is fixed at two samples per second

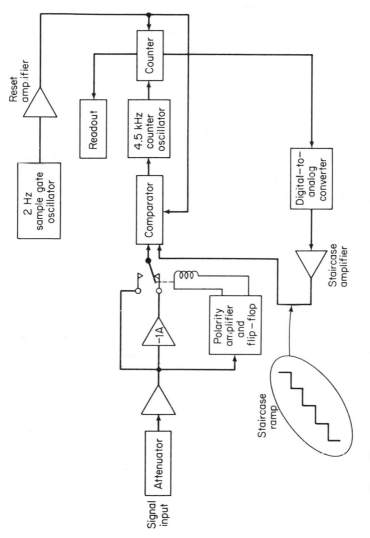

Fig. 1-39   Basic staircase-ramp digital voltmeter circuit. (Courtesy Hewlett-Packard.)

by the 2-Hz sample oscillator) a reset pulse resets the staircase ramp to zero, and the measurement starts over.

Integrating-type Digital Meter

One of the problems with a ramp digital meter is that measurements at the end of the timer interval can occur simultaneously with a noise burst (where direct current must be measured in the presence of other signals). These noise signals near the second coincidence of the ramp voltage could lengthen (or shorten) the time interval, thus making the count incorrect. This problem can be overcome by means of an integrating digital meter that makes the measurement on the basis of *voltage-to-frequency* conversion, rather than voltage-to-time, as used in the ramp meter.

A typical voltage-to-frequency integrating converter is shown in Fig. 1-40. The circuitry functions as a feedback-control system that governs the rate of pulse generation, making the average voltage of the rectangular pulse train equal to the d-c input voltage.

As shown, a positive voltage at the input results in a negative-going ramp at the output of the integrator. The ramp continues until it reaches a voltage level that triggers the level detector. In turn, the level detector triggers a pulse generator. The pulse generator produces a rectangular pulse with closely controlled width and amplitude just sufficient to draw enough charge from capacitor C to bring the input of the integrator back to the starting level. The cycle then repeats.

The ramp slope is proportional to the input voltage. (For example, a higher voltage at the input results in a steeper slope, resulting in a shorter time duration for the ramp.) As a result, the pulse repetition rate will be

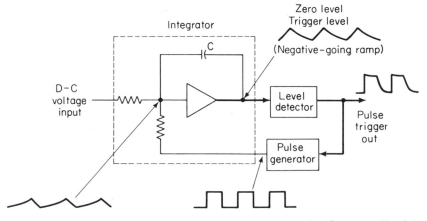

Fig. 1-40   Typical voltage-to-frequency conversion circuit. (Courtesy Hewlett-Packard.)

higher. Since the pulse repetition rate is proportional to the input voltage, the pulses can be counted during a known time interval to find a digital measure of the input voltage.

Although a voltage ramp is generated in this type of DVM, the amplitude is only a fraction of a volt, and the accuracy of the analog-to-digital conversion is determined not only by the characteristics of the ramp but also by the area of the feedback pulses.

The primary advantage of this type of analog-to-digital conversion is that the input is "integrated" over the sampling interval, and the reading represents a *true average* of the input voltage. The pulse repetition frequency "tracks" a slowly varying input voltage so closely that changes in the input voltage are accurately reflected as changes in the pulse repetition rate. The total pulse count during a sampling interval therefore represents the average frequency and thus the average voltage. This is important when noisy signals or voltages are encountered, since the noise can be averaged out during the measurement.

Another advantage of the integrating circuit is that the measurement circuit can be completely isolated (by shielding and transformer coupling) from the counter/readout circuits. This technique is shown in Fig. 1-41.

Integrating/Potentiometric-type Digital Meter

An integrating/potentiometric meter combines the continual measurement of true average input voltage with accuracy from precision resistance ratios and stable reference voltages. A block diagram of a typical integrating/potentiometric meter is shown in Fig. 1-42. Note that the instrument is divided into three sections: a voltage-to-frequency (V/F) converter, a counter, and a digital-to-analog (D/A) converter.

As in the case of the conventional integrating meter, the voltage-to-frequency converter generates a pulse train with a rate exactly proportional to the input voltage. This pulse train is gated for a precise time interval and

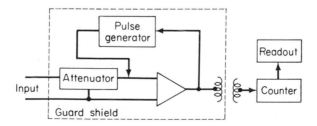

Fig. 1-41   Simplified block diagram of guard-circuit technique. (John D. Lenk, Handbook of Electronic Test Equipment, © 1971, p. 58. Courtesy Prentice-Hall.)

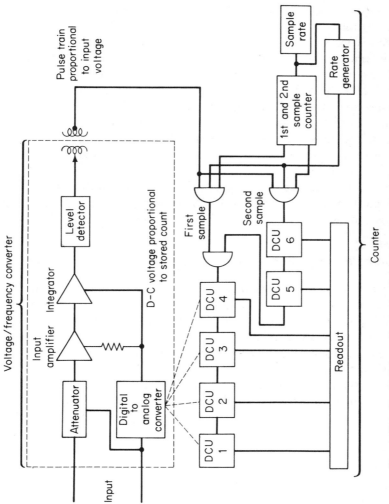

Fig. 1-42 Basic integrating/potentiometric type digital voltmeter. (Courtesy Hew-lett-Packard.)

is fed to the first four places in a six-digit counter. The stored (undisplayed) count is transferred to the D/A converter, which produces a highly accurate d-c voltage proportional to the stored count. This voltage is subtracted from the unknown voltage at the input to the V/F converter.

The pulse drain from the V/F converter is again gated, this time to the last two places in the six-digit counter. At the end of the second gate period, the total count is transferred to the six corresponding readouts. The counter display or readout thus indicates the integral of the input voltage. Accuracy of this instrument is dependent on the accuracy of the D/A converter, as well as the V/F converter.

Dual-Slope Integration-type Digital Meter

The dual-slope-type meter uses an entirely different principle to measure d-c voltage (or other values, such as resistance or a-c voltage). A block diagram of a dual-slope meter is shown in Fig. 1-43. This instrument measures d-c voltage by use of an integrator that produces a time interval proportional to the average value of the applied d-c voltage. The time interval determines the gate time of the counter and thus the number of pulses totalized. In this way, the number of pulses is proportional to the average of the d-c voltage measured.

During a precisely controlled time period (1/10 or 1/60 second, as selected by a front-panel control) the integrator charges up to a value proportional to the average value of the d-c input voltage. This charging voltage is the "up-slope" of the integrator output. After a time period, a precise reference voltage of opposite polarity is switched to discharge the integrator. This discharge is the "down-slope" of the integrator. The zero crossing of the output voltage is detected by a zero detector circuit. The counter is enabled to totalize pulses from the crystal oscillator during the discharge time or down-slope of the integrator. Since the discharge time is proportional to the stored voltage, the number of pulses totalized is proportional to the input voltage.

After completion of the integration cycle, the input amplifier is disconnected and automatically zeroed before the next measurement is taken. This autozeroing effectively compensates for d-c drift and eliminates the need for a chopper amplifier and front-panel zero controls.

Alternating-Current Voltage Measurements with Dual-Slope Meter

As shown in dashed lines on the block of Fig. 1-43, the dual-slope meter can be used to measure a-c voltage as well as d-c voltage. However, this requires that the ac be converted to dc before application to the integrator. Either average or RMS values of the ac can be measured.

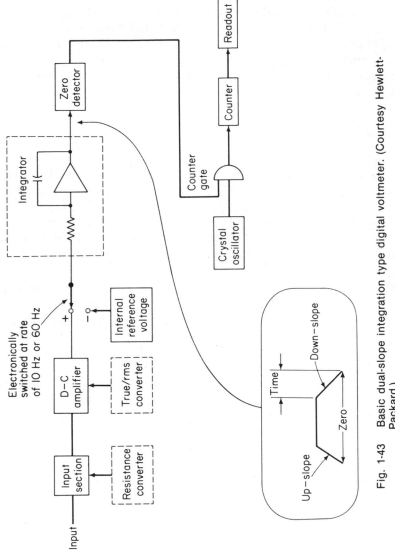

Fig. 1-43   Basic dual-slope integration type digital voltmeter. (Courtesy Hewlett-Packard.)

The average-value converter circuit is shown in Fig. 1-44. Note that this circuit is similar to the average-responding voltmeter. This negative feedback limits the output to 1 V dc, no matter what the a-c input value. Typically, the circuit produces a d-c output voltage between 0 and 1 V, proportional to the average value of the applied a-c voltage.

The true RMS converter is shown in Fig. 1-45. The input circuitry consists of an operational amplifier whose gain is accurately controlled to achieve attenuation of the input signal. An a-c output from the input amplifier is sent to the modulator, and a second output is used as a trigger for the sync-generator. The sync-generator produces a 5-Hz square wave and is used to synchronize the modulator and demodulator to the input signal.

A 1-kHz oscillator drives the d-c-to-square-wave converter that converts the d-c output of the a-c converter to a reference square wave. The amplitude of the square wave is proportional to the d-c output. The output of the modulator, at a nominal 5 Hz, consists of a composite signal made up of one-half input signal and one-half reference square wave (as shown in Fig. 1-45).

The automatic gain control (AGC) amplifier controls the gain of the sampling amplifier and the integrator. This keeps the RMS value of the signal applied to the thermocouple constant, and holds the gain of the system constant, regardless of the level of the input signal. The output of the thermocouple varies between the RMS value of the input and reference signals. This error signal is amplified, and two 180° out-of-phase signals are sent to the demodulator. The demodulator acts as full-wave rectifier. The output pulses are amplified and integrated to develop the positive d-c voltage output. This d-c voltage is continuously corrected at nominally 5

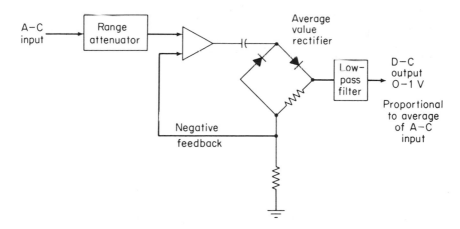

Fig. 1-44   Basic average-value a-c/d-c converter. (John D. Lenk, Handbook of Electronic Test Equipment, © 1971, p. 61. Courtesy Prentice-Hall.)

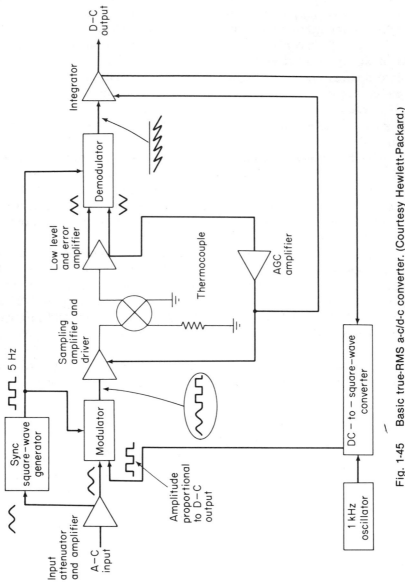

Fig. 1-45 Basic true-RMS a-c/d-c converter. (Courtesy Hewlett-Packard.)

times/sec (set by the 5-Hz sync-square-wave generator) to ensure a d-c voltage output proportional to the RMS value of the input signal.

Resistance Measurements with Dual-Slope Meter

As shown by dashed lines in Fig. 1-45, the dual-slope meter can also be used to measure resistance. However, this requires that the resistance be converted to direct current before application to the integrator.

The resistance converter circuit is shown in Fig. 1-46. The resistance measurements are made by feeding a constant current through the unknown resistor and measuring the resultant voltage across the resistor. The current source supplies three constant currents of 1 ma, 10 μA, and 1μA, and an open-loop voltage of 17 V maximum. The current is held constant by means of an operational amplifier and feedback resistors. To increase measurement accuracy, the internal reference voltage (Fig. 1-43) is disabled, and the ohms reference voltage (Fig. 1-46) is used to discharge the integrator.

Counter/Readout and Divider Circuit Operation

Counter circuits found in most digital meters use *decade counters* (also called *decades*) made up of four *binary* counters that convert the count to a BCD (binary-coded decimal) code. Digital meter counter circuits also include *decoders* for conversion of the code to decimal form and *readouts* that display the information directly in decimal form. The same circuit used for decade counters (four binary counters) are also used as *dividers*. Therefore,

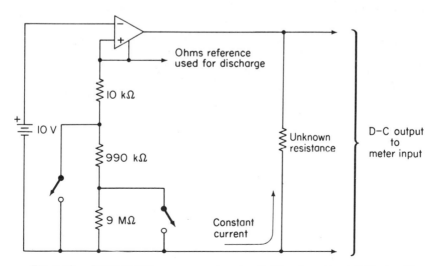

Fig. 1-46   Basic resistance converter circuit. (Courtesy Hewlett-Packard.)

it is necessary to understand operation of the basic decade before going into how the decade is used in digital meters.

Decades Decade counters (or decades) serve two purposes in the counter circuits of digital meters. First, the decade divides frequencies by 10. That is, the decade produces one output for each 10 input pulses or signals. This permits several frequencies to be obtained from one basic frequency. For example, a 1-MHz time base (such as the crystal oscillator shown in Fig. 1-43) can be divided to 100 kHz by one decade divider, to 10 kHz by two decade dividers, to 1 kHz by three decade dividers, and so on. When decades are used for division, they are sometimes called *scalers*, although "dividers" is a better term. The second purpose of a decade is to convert a count to a BCD code. The division function of a decade is discussed here first.

The basic unit of a decade divider is a 2:1 scaler, called a *binary counter*. This unit uses a *bistable multivibrator* or a logic *flip-flop* (FF), depending on design. A basic flip-flop is shown in Fig. 1-47. Although the circuit shown uses discrete components, decade flip-flops are most often found in IC form, with all four FFs in one package. Sometimes several decades are found in one package.

The flip-flop of Fig. 1-47 is made up of two cross-coupled AND gates. Such a FF has two stable states, with the gate *A* output positive and the gate *B* output negative, and vice versa. The first input pulse flips the circuit from one state to the other. The second input pulse flops the circuit back to the original state, hence the name "flip-flop." Each time the circuit is flipped from one state to the other and back again (requiring two input pulses) a single (complete) output pulse is produced. The output from the FF may be taken from either gate *A* or gate *B*. The important point to remember is that the outputs are always in opposite states, and that it takes two input pulses to produce one output pulse, at either output.

The output pulses of one FF may be applied to the input of another similar FF for further frequency division. This is called *cascading*. A basic binary counter uses a cascaded chain of four FFs, as shown in Fig. 1-48.

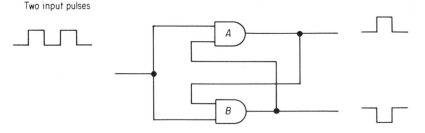

Fig. 1-47   Basic flip-flop (binary counter or 2:1 scaler).

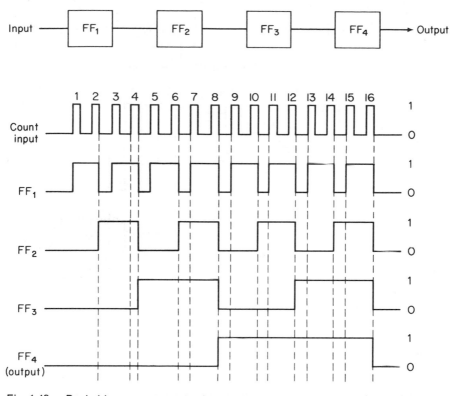

Fig. 1-48   Basic binary counter with four flip-flops in cascade. (John D. Lenk, Handbook of Electronic Test Equipment, © 1971, p. 242. Courtesy Prentice-Hall.)

The count of this chain is 16. That is, for every 16 input pulses to be counted, the output of FF1 is 8, of FF2 is 4, of FF3 is 2, and of FF4 is 1. When the count is to be divided by 10, some of the FF outputs are fed back to cancel the pulses, as shown in Fig. 1-49. Here, for every 10 input pulses to be counted, the output of FF1 is 5, of FF2 is 3, of FF3 is 2, and of FF4 is 1.

Decade-to-BCD Conversion The decades shown in Figs. 1-47 through 1-49 are used for division. The same basic circuit can be used to convert a series of pulses into a binary code such as BCD. To understand this function, it is necessary to understand the binary counting system and how the BCD code relates to this system. We will review these subjects before discussing decade-to-BCD conversion.

In the binary system, all numbers can be made up by using only ones and zeros (1 and 0), rather than zero through nine, as in the familiar decimal system. Consequently, instead of requiring 10 different values to represent one digit, circuits using the binary method need only two values for each

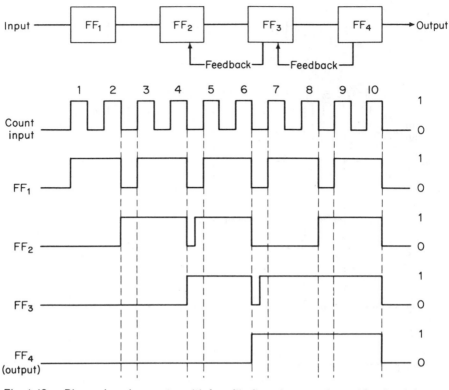

Fig. 1-49 Binary decade counter with four flip-flops in cascade, and feedback from FF₃ to FF₂, FF₄ to FF₃. (John D. Lenk, Handbook of Electronic Test Equipment, © 1971, p. 243. Courtesy Prentice-Hall.)

digit. In counters, and other digital electronic equipment, these values are easily indicated by the presence or absence of a signal (or pulse), or by positive and negative signals, or even by two different voltage levels. Thus, binary lends itself to any electronic device.

In binary, the value of each digit is based on 2 and the powers of 2. In binary number, the extreme right-hand digit is multiplied by 1, the second-from-the-right digit is multiplied by 2, the third-from-the-right digit is multiplied by 4, and so on. This can be displayed as follows:

$$2^8 \quad 2^7 \quad 2^6 \quad 2^5 \quad 2^4 \quad 2^3 \quad 2^2 \quad 2^1 \quad 2^0$$
$$256 \quad 128 \quad 64 \quad 32 \quad 16 \quad 8 \quad 4 \quad 2 \quad 1$$

In binary, if the digit is zero, the value is zero. If the digital is 1, the value is determined by the position from the right. For example, to represent the number 77 in binary form, the following combination of zeros and ones is used:

256	128	64	32	16	8	4	2	1
0	0	1	0	0	1	1	0	1
		64+	0 +	0 +	8+	4+	0+	1= 77

which means 1001101 in pure binary form = 77.

The binary-coded-decimal (BCD) system combines the advantages of the binary system (the need in digital electronic circuits for only two states, one or zero) and the convenience of the familiar decimal representation. In the BCD system, a number is expressed in normal decimal coding, but each digit in the number is expressed in binary form.

For example, the number 37 in BCD form appears as

	Tens Digit	Units Digit
Decimal	3	7
BCD	0011	0111

Note that 4 bits of information are needed for each digit. In general, 4 bits yield 16 possible combinations. However, in BCD only 10 combinations are needed.

Figure 1-50 shows a decade circuit capable of converting a series of pulses into the BCD code. One FF is used for each of the digits. Input pulses are fed to the 1-FF, the 2-, 4-, and 8-FFs are cascaded and receive pulses after the 1-FF.

At the beginning of a count, the 8421 lines (representing the 8-, 4-, 2-, and 1-FF outputs, respectively) are at zero. In a typical counter, the decades are set (or reset) to this condition by the application of a voltage or pulse produced by a reset button, or by a sample rate generator, or by pulses from the decades at the end of a 10 count.

When the first pulse in the count is applied, the 1-FF changes stages. The 1-line changes from 0 to 1. When the second pulse is applied, the 1-FF again changes states. The 1-line goes from 1 to 0, and the 2-line goes from 0 to 1.

With the third pulse applied, the 1-FF goes from 0 to 1, but the 2-FF remains at 1. Remember that the 2-FF will change states for each complete cycle (or two-state change) of the 1-FF.

When the fourth pulse is applied, the 1-FF goes to 0, as does the 2-FF. This causes the 4-FF to change states (the 4-line goes to 1).

This process is repeated until a nine count is reached. At that point, the 8-FF moves from 1 to 0. This 0 output is returned to the reset line and

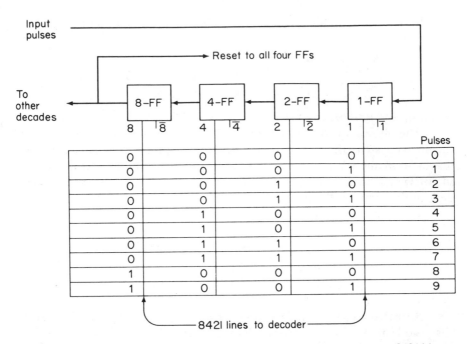

Input
pulses

Reset to all four FFs

To
other
decades

8	$\overline{8}$	4	$\overline{4}$	2	$\overline{2}$	1	$\overline{1}$	
								Pulses
0		0		0		0		0
0		0		0		1		1
0		0		1		0		2
0		0		1		1		3
0		1		0		0		4
0		1		0		1		5
0		1		1		0		6
0		1		1		1		7
1		0		0		0		8
1		0		0		1		9

8421 lines to decoder

Fig. 1-50   Binary decade circuit for converting a series of pulses into an 8421 binary code. (John D. Lenk, Handbook of Electronic Test Equipment Test Equipment, © 1971, p. 244. Courtesy Prentice-Hall.)

serves to reset all the FFs to the negative or 0 state. The output from the 8-FF can also be applied to the 1-FF of another decade. Any number of decades can be so connected. One decade is required for each readout numeral (in most cases).

BCD-to-Seven-Segment Readout Conversion   Most electronic readouts use some form of seven-segment display, as shown in Fig. 1-51. The numerals (from 0 through 9) are formed when the corresponding segments are turned on. For example, to form the numeral 8, all segments are turned on simultaneously. To form the numeral 3, all segments, *except e and f*, are turned on. The method of turning on the segments depends on the type of display used. The most common forms of numerical readouts are liquid crystal displays (LCD), light-emitting diode (LED), gas-discharge, fluorescent, and incandescent. Most digital meters use either LCD or LED.

The liquid crystals used in LCDs are fluids that flow like a liquid but which have some of the optical characteristics of solid crystals. LCDs consist of certain organic compounds whose characteristics change state when placed in an electric fluid. Thus, images can be created according to predetermined patterns (segments in this case). Since no light is emitted or

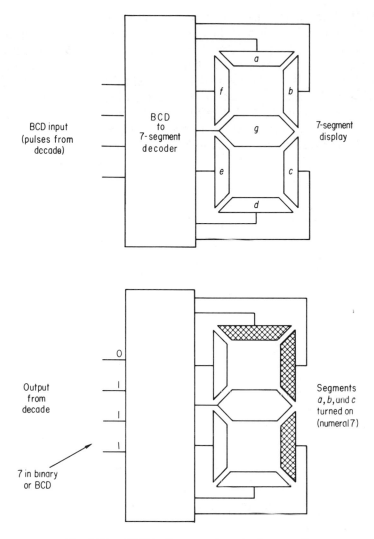

Fig. 1-51   BCD-to-7-segment readout conversion.

generated, very little power is required to operate LCDs, and they are well suited for battery operation. LCDs have good readability in sunlight or bright light. However, in low-ambient-light conditions, some form of light source (either within or external to the display) is generally used.

LEDs are semiconductor junction diodes which produce light when forward-biased (anode more positive than cathode). The semiconductor material is either gallium arsenide phosphide (GaAsP) or gallium phosphide (GaP), with the former being more prevalent in red display applications.

With either LCD or LED readouts, the individual segments are turned on when a voltage is applied to the selected segment. In the readout of Fig. 1-51, the voltage is applied by a BCD-to-seven-segment decorder, in response to pulses from a decade such as that shown in Fig. 1-50. For example, when seven pulses to be counted are applied to the decade of Fig. 1-50, the output of the decade to the decoder is 0111, which is seven in binary or BCD. The decoder of Fig. 1-51 then converts this to voltages on segments a, b, and c, with all other segments receiving no voltage (turned off). Segments a, b, and c are turned on, and the numeral 7 is formed.

2

Accessory Probes

Except for certain special-purpose laboratory instruments, most meters can perform their functions without accessories. One exception to this is the need for a test probe. To get the full benefit from any type of probe, the user needs a working knowledge of how the probe performs its function, even though the actual operating procedure is quite simple. Therefore, this chapter describes the probes most commonly used with modern meters and how probes can be made from simple parts. Such probes can be used to extend the capabilities of shop-type meters.

2-1 THE BASIC METER PROBE

In its simplest form, the basic meter probe is a "test prod." In physical appearance, the probe is a thin metal rod connected to the meter input through an insulated flexible lead. All but a small tip of the rod is covered with an insulating handle, so that the probe can be connected to any point of a circuit without touching nearby circuit parts. Sometimes the probe tip is provided with an alligator clip so that it is not necessary to hold the probe at the circuit point.

Such probes work well on circuits carrying d-c and audio-frequency signals. If, however, the ac is at a high frequency or if the gain of a meter amplifier is high, it may be necessary to use a special low-capacitance probe. Hand capacitance in a simple probe can cause hum pickup, particularly if the amplifier gain is high. This condition can be offset by shielding in a low-capacitance probe. In a more important problem, the input impedance of the meter is connected directly to the circuit under test by a simple probe and may change circuit conditions. The low-capacitance probe contains a series capacitor and resistor that increase the meter impedance.

2-2 LOW-CAPACITANCE PROBES

The basic circuit of a low-capacitance probe is shown in Fig. 2-1. The series resistance R_1 and capacitance C_1, as well as the parallel or shunt resistance R_2, are surrounded by a shielded handle. The values of R_1 and C_1 are preset at the factory by screwdriver adjustment.

In most low-capacitance probes, the values of R_1 and R_2 are selected so that they form a 10 : 1 voltage divider between the circuit under test and the meter input. The operator must remember that voltage indications will be one-tenth of the actual value when the probe is used.

The capacitance value of C_1, in combination with the values of R_1 and R_2, also provides a capacitance reduction (usually in the range of 3 : 1 to 11 : 1). R_1 and C_1 are usually factory-adjusted and *should not* be disturbed unless recalibration is required. (Calibration is discussed further in Chapter 4.)

There are probes which combine the features of low-capacitance probes and the simple probes described in Sec. 2-1. In such probes, a switch (shown

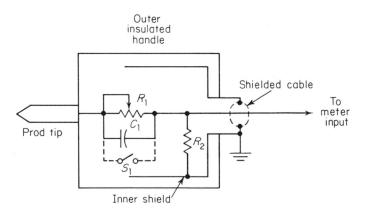

Fig. 2-1   Typical low-capacitance probe circuit.

as S_1 in dashed form in Fig. 2-1) is used to short both C_1 and R_1 when a direct input is required. With S_1 open, both C_1 and R_1 are connected in series with the input, and the probe provides the low-capacitance feature.

2-3 RESISTANCE-TYPE VOLTAGE-DIVIDER PROBES

A resistance-type voltage-divider probe is used when the primary concern is a reduction of voltage. The resistance-type probe, shown in Fig. 2-2, is similar to the low-capacitance probe described in Sec. 2-2, except that the frequency-compensating capacitor is omitted. Usually, the straight resistance-type probe is used when a voltage reduction of 100 : 1 or greater is required and when a flat frequency response is of no particular concern.

As shown in Fig. 2-2, the values of R_1 and R_2 are selected to provide the necessary voltage division and to match the input impedance of the meter. Resistor R_1 is usually made variable so that an exact voltage division can be obtained.

Because of their voltage-reduction capabilities, resistance-type probes are often known as *high-voltage* probes. Some resistance-type probes are capable of measuring potentials at or near 40 kV (with a 1000 : 1 voltage reduction).

2-4 CAPACITANCE-TYPE VOLTAGE-DIVIDER PROBES

In certain isolated cases, the resistance-type voltage-divider probes described in Sec. 2-3 are not suitable for measurement of high voltages because stray

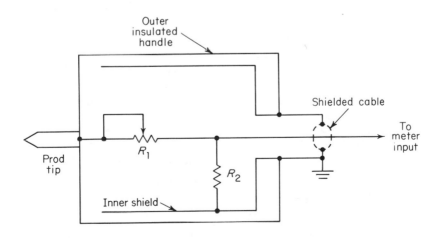

Fig. 2-2   Typical resistance-type voltage-divider probe circuit.

conduction paths are set up by the resistors. A capacitance-type probe, shown in Fig. 2-3, can be used in those cases.

In such capacitance probes, the values of C_1 and C_2 are selected to provide the necessary voltage division and to match the input capacitance of the meter. Capacitor C_1 is usually made variable so that an exact voltage division can be obtained.

2-5 RADIO-FREQUENCY PROBES

When the signals to be measured by a meter are at radio frequencies and are beyond the frequency capabilities of the meter circuits, a radio-frequency or RF probe is required. Such probes convert (rectify) the RF signals into a d-c output voltage that is equal (almost) to the peak RF voltage. The d-c output of the probe is then applied to the meter input and is displayed as a voltage readout in the normal manner. In some RF probes, the meter reads peak RF voltage, whereas in other probes the readout is in RMS voltage.

The basic circuit of a radio-frequency probe is shown in Fig. 2-4. This circuit can be used to provide either peak output or RMS output. Capacitor C_1 is a high-capacitance, d-c blocking capacitor used to protect diode CR_1. Usually, a germanium diode is used for CR_1, which rectifies the RF voltage and produces a d-c output across R_1. In some probes R_1 is omitted so that the d-c voltage is developed directly across the input circuit of the meter. This d-c voltage is equal to the peak RF voltage less whatever forward drop exists across the diode CR_1.

When it is desired to produce a d-c output voltage equal to the RMS of the RF voltage, a series-dropping resistor (shown in dashed form in Fig. 2-4

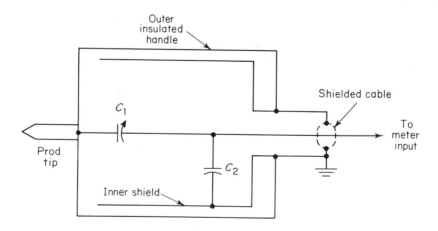

Fig. 2-3　Typical capacitance-type voltage-divider probe circuit.

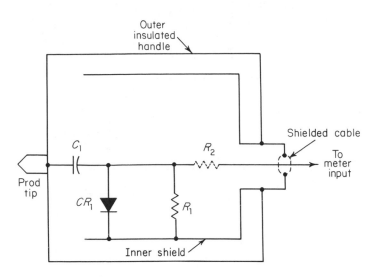

Fig. 2-4   Typical radio-frequency probe circuit.

at R_2) is added to the circuit. Resistor R_2 drops the d-c output voltage to a level that equals 0.707 of the peak RF value.

2-6 DEMODULATOR PROBES

The circuit of a demodulator probe is essentially like that of the RF probe described in Sec. 2-5. The circuit values and the basic functions are somewhat different.

The prime purpose of a demodulator probe is demodulating an amplitude-modulated signal and converting the modulation envelope (low-frequency component such as audio) into a d-c output voltage.

The basic circuit of the demodulator probe is shown in Fig. 2-5. Here, capacitor C_1 is a low-capacitance, d-c blocking capacitor. (In the RF probe a high-capacitance value is required for C_1 to ensure that the diode operates at the peak of the RF signal. This is not required for a demodulator probe.)

Germanium diode CR_1 demodulates (or detects) the amplitude-modulated signal and produces a voltage across load resistor R_1. (The load resistor can be omitted in some cases.) The voltage developed across R_1 is pulsating dc and is proportional in amplitude to the modulating voltage. The voltage has the same approximate waveform and frequency as the modulating voltage. Resistor R_2 is used primarily for isolation between the circuit under test and the meter input. In general, demodulator probes are used primarily for *signal tracing*, and their output is not calibrated to any particular value (peak, RMS, etc.).

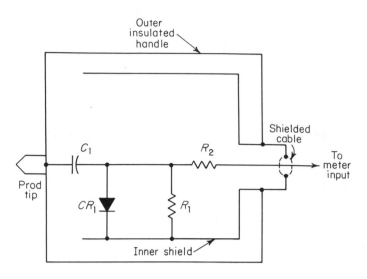

Fig. 2-5   Typical demodulator probe circuit.

2-7   USING PROBES EFFECTIVELY

The proper use of probes is important when reliable measurements are to be made. The following notes summarize some of the more important considerations in the use and adjustment of probes. Probe-use techniques for specific measurements are discussed further in other chapters and sections.

Probe Compensation

The capacitors that compensate for excessive attenuation of high-frequency signal components (through the probe-resistance dividers) affect the entire frequency range. These capacitors must be adjusted so that the higher-frequency components will be attenuated by the same amount as low frequencies and direct current. Adjustment of the compensating capacitors should be done at the factory with the proper test equipment. Probe compensation is discussed further in Chapter 4.

Circuit Loading

Connection of a meter to a circuit may alter the amplitude or waveform at the point of connection. To prevent this, the impedance of the circuit being measured must be a small fraction of the input impedance of the meter. When a probe is used, the *probe's impedance* determines the amount of circuit loading.

The ratio of the two impedances (probe and circuit) represents the amount of probable error. For example, a ratio of 100 : 1 (perhaps a 100-megohm probe to measure the voltage across a 1-megohm circuit) will account for about a 1 percent error. A ratio of 10 : 1 would produce a possible 9 percent error.

Remember that the input impedance is not the same at all frequencies but continues to get smaller at higher frequencies because of the input capacitance. (Capacitive reactance and impedance decrease with an increase in frequency.) All probes will have some input capacitance. Even an increase at audio frequencies may produce a significant change in impedance. When using a shielded cable with a probe to minimize pickup of stray signals and hum, the additional capacitance of the cable should be recognized. The capacitance effects of a shielded cable can be minimized by terminating the cable at one end in its characteristic impedance. Unfortunately, this is not always possible with most meters.

The reduction of resistive loading due to probes may not be as much as the attenuation ratio of the probe, but capacitive loading will not be reduced to the same extent because of the additional capacitance of the probe cable. For example, a typical 5 : 1 attenuator probe may be able to reduce capacitive loading somewhat better than 2 : 1. A 50 : 1 attenuator probe may reduce capacitive loading by about 10 : 1. Beyond this point, little improvement can be expected because of stray capacitance at the probe tip.

Practical Uses of Probes

Whether a particular probe connection is disturbing a circuit can be judged by attaching and detaching another connection of similar kind (such as another probe) and observing any difference in the meter reading. If there is little or no change when the additional probe is touched to the circuit, it is safe to assume that the meter probe has little effect on the circuit.

Long probes should be restricted to the measurement of relatively slow changing signals (dc and low-frequency ac). The same is true for long ground leads.

The ground lead should be connected to a point where no hum or high-frequency signal components exist in the ground path between that point and the signal pick-off point.

Reliable measurements involving frequency components higher than 10 MHz require probes with special inner conductors.

A 10-ohm resistor at the tip of a probe may prevent "ringing" of the ground lead when the probe is connected to very low impedance signal sources having very high frequency components.

Resistive loading may be eliminated entirely by using a small (0.002

μF) capacitor in series with the probe tip, at the sacrifice of some low-frequency response.

Avoid applying more than the rated peak voltage to a probe. Using a high-voltage coupling capacitor between a probe tip and a very high d-c level *may not always prevent probe burnout.* This is because the capacitor must charge and discharge *through the probe.* If care is taken to charge and discharge the blocking capacitor through a path that shunts the probe, the technique can be successful. One way to accomplish this is to ground *the junction of the capacitor and the probe tip* whenever the capacitor is being charged or discharged.

For example, assume that the probe is to be used to measure 500 V and that the probe is rated at 250 V. By connecting a high-voltage capacitor to the probe tip, grounding the junction of the probe tip and capacitor, and then connecting the free end of the capacitor to the 500-V point, the capacitor will be charged and the measurement can continue as normal. At the completion of the measurement, the capacitor should be discharged by touching the probe-capacitor junction to ground. In this way, both charging and discharging of the capacitor will be accomplished outside the probe.

Be certain to check for proper probe compensation whenever changing a probe or when making an important measurement.

2-8 PROBE CIRCUITS FOR THE EXPERIMENTER

Most of the circuits used in probes are quite simple and can be made up by experimenter to extend the range of their meters. Usually, these homemade probes are best used for signal tracing or for *relative signal level* checks rather than for precision voltage measurements. However, it is possible to calibrate the probe circuits with a particular meter to provide measurements accurate enough for all but precision laboratory requirements.

If a probe is available as an accessory for a particular meter, *that probe* should be used in favor of any homemade probe. The manufacturer's probe will be matched to the meter in calibration, frequency compensation, and so on.

If a probe is not available for a particular meter or if an emergency probe must be made up, the following notes will permit the student or technician to adapt a probe circuit to his or her particular meter.

Half-Wave Probe

The half-wave probe will provide an output to the meter that is (approximately) equivalent to the *peak* value of the voltage being measured. A practical half-wave signal-tracing probe circuit is shown in Fig. 2-6. Since most

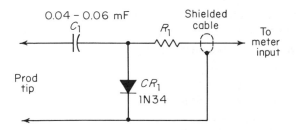

Fig. 2-6   Half-wave signal-tracing probe circuit.

shop-type meters are calibrated to read in RMS values, the probe output must be reduced to 0.707 of the peak value. This is accomplished by selecting the correct value for resistor R_1.

The value of R_1 could be determined by calculation. But for practical purposes, a variable resistor should be used during calibration. Then a fixed resistor of the same value can be used for the actual probe circuit. The following steps describe the calibration and fabrication procedure:

1. Connect the probe components (coupling or blocking capacitor C_1, diode CR_1, and calibrating resistor R_1) to a voltage source and to the meter.

2. Set the meter to measure d-c voltage. The meter can be either a VOM or an electronic meter. Best results will be obtained with a higher ohms-per-volt meter, particularly with low-frequency responses. This is because little current is required for the charging and discharging of capacitor C_1 with a high resistance circuit. Therefore, an electronic meter will work better than a VOM. Likewise, a 100,000-ohms/V VOM will produce better low-frequency response than will a 20,000-ohms/V VOM.

3. Adjust the voltage source to some precise value, such as 10 V peak.

4. Adjust the calibrating resistor R_1 until the meter indicates 7.07 V RMS (assuming a 10-V peak source is used).

5. Remove the voltage source and disconnect the variable resistor from the circuit. Measure the value of the variable resistor, select a fixed resistor of the same value, and connect it into the probe circuit.

6. Reapply the voltage source (10 V peak) and check that the meter indication is 7.07 V RMS. If so, the probe circuit is complete and can be put into a suitable package.

7. Usually, a value of 10 to 20 kilohms for R_1 will be satisfactory with most VOM circuits. A much higher value (in the order of 1 megohm) will be required for most electronic meter circuits. Also, many electronic meter circuits have an *input cable* (shielded with a probe tip) rather than the simplest test leads and prods of a VOM. The input

cable may contain an isolating resistor. If so, the value of R_1 will have to be somewhat near the value of the isolating resistor.

8. Make certain that the meter is set to measure dc, since the probe output is direct current.

9. 1N34A diode is shown for CR_1. Any type of small-signal diode can be used. The maximum voltage value with which the probe can be used is limited by the peak-to-peak maximum of diode CR_1.

10. A value of 0.04 to 0.06 μF is shown for capacitor C_1. The following points must be considered in selecting the value of C_1. A high value for C_1 will give better low-frequency response. On the other hand, a high value for C_1 could cause damage to diode CR_1 if the probe were connected to a high-voltage source and if C_1 were charged through CR_1. The 0.04- to 0.06μF value represents a compromise between high-current surge protection for CR_1 and good low-frequency response.

11. Once the probe has been made up, it should be checked for frequency response as described in Chapter 4.

Full-Wave Probe

The full-wave probe will provide an output to the meter that is (approximately) equivalent to the *peak-to-peak* value of the voltage being measured. This is particularly important when measuring pulses, square waves, and any other complex waveform.

A practical full-wave signal-tracing probe circuit is shown in Fig. 2-7. Since most shop-type meters are calibrated to read in RMS values, the probe output must be reduced to 0.3535 of the peak-to-peak value. This is accomplished by selecting the correct value for resistor R_1.

Some commercial electronic meters are provided with peak-to-peak

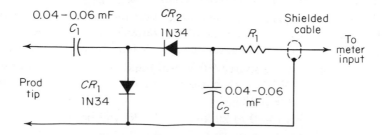

Fig. 2-7   Full-wave signal-tracing probe circuit.

scales. These meters are used in certain laboratory applications and in TV service to measure such values as horizontal-oscillator or deflection-coil voltages, input to the video amplifier, and output of the vertical amplifier. Commercial peak-to-peak meters are provided with their own full-wave probes or input circuits. The following steps describe how a conventional electronic meter can be adapted to measure peak-to-peak voltages, using the RMS scales. As in the case of the half-wave probe, a temporary variable resistance is used to obtain the correct value for calibrating resistor R_1. In the final circuit, the variable resistance is replaced by a fixed resistance of the correct value.

1. Connect the probe components (capacitors C_1 and C_2, diodes CR_1 and CR_2, and calibrating resistor R_1) to a voltage source and to the meter.

2. Set the meter to measure d-c voltage. An electronic meter will produce the best results, although a very sensitive VOM (100,000 ohms/V or higher) may prove satisfactory.

3. Adjust the voltage source to some precise value, such as 10 V peak to peak.

4. Adjust the calibrating resistor R_1 until the meter indicates 3.535 V RMS (assuming that a 10-V peak-to-peak source is used).

5. Remove the voltage source and disconnect the variable resistor from the circuit. Measure the value of the variable resistor, select a fixed resistor of the same value, and connect it into the probe circuit.

6. Reapply the voltage source (10 V peak to peak) and check that the meter indication is 3.535 V RMS. If so, the probe circuit is complete and can be put into a suitable package.

7. Usually, a value of 1 megohm or greater will be required for a typical electronic meter circuit.

8. Make certain that the meter is set to measure dc since the probe output is direct current.

9. 1N34A diodes are shown for CR_1 and CR_2. Any type of small-signal diode can be used. Best results will be obtained when the two diodes are matched in forward voltage drop, reverse resistance, and so on. The maximum voltage value with which the probe can be used is limited by the peak-to-peak maximum value of either CR_1 or CR_2.

10. As in the half-wave probe, a value of 0.04 to 0.06 μF is shown for capacitors C_1 and C_2.

11. Once the probe has been made up, it should be checked for frequency response as described in Chapter 4.

Demodulator Probe

Both the half-wave and full-wave probes are used to convert high-frequency signals (usually an RF carrier) into a d-c voltage that can be measured on a meter. When the high-frequency signals contain ac or pulsating dc modulation (such as a modulated RF carrier), a demodulator probe will be more effective for signal tracing.

A practical demodulator probe circuit is shown in Fig. 2-8. Such a probe is most effective with an electronic meter and is well suited to signal-tracing use. The demodulator probe is similar to the half-wave probe of Fig. 2-6, except for the low capacitance value of C_1 and the parallel resistor R_2. These two components act as a filter. The demodulator probe produces both an a-c and a d-c output. The RF carrier frequency is converted into a d-c voltage equal to the peak of the RF carrier. The low-frequency modulating voltage appears as ac at the probe output.

In use, the electronic meter is set to measure dc, and the RF carrier is measured. Then the meter is set to ac and the modulating voltage is measured. The calibrating resistor R_1 is adjusted so that the d-c scale reads the RMS value. (The d-c output of the probe is actually the peak value of the RF carrier.) The modulating voltage measurement will be RMS or peak to peak, depending upon the scale of the electronic meter. The following steps describe the calibration procedure.

1. Connect the probe components (capacitor C_1, diode CR , and resistors R_1 and R_2) to a voltage source and to the meter.
2. Set the meter to measure d-c voltage.
3. Adjust the voltage source to some precise value, such as 10 V peak. Use an unmodulated source for the initial calibration.
4. Adjust the calibrating resistor R_1 until the meter indicates 7.07 V RMS (assuming that a 10-V peak source is used.)
5. Remove the voltage source and disconnect the variable resistor from the circuit. Measure the value of the variable resistor. Select a fixed resistor of the same value and connect it into the probe circuit.

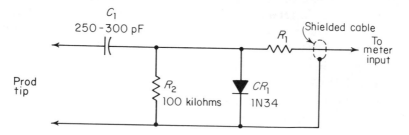

Fig. 2-8   Demodulator probe circuit.

6. Reapply the voltage source (10 V peak) and check that the meter indication is 7.07 V RMS. (Actually, the d-c scale of the meter is arbitrarily used as an RMS scale.)

7. Usually, a 0.25-megohm value for R_1 will be satisfactory. If the value of R_1 is increased beyond 0.25 megohm, there is a tendency to attenuate both the d-c and a-c outputs.

8. Make certain that the meter is set to measure dc when the RF carrier is to be monitored and to measure ac when the modulating voltage is to be read out.

9. A 1N34A is shown for CR_1. Again, any type of small-signal diode can be used. The maximum voltage value with which the probe can be used is limited by the peak-to-peak maximum of the diode CR_1.

10. A value of 250 to 300 pF is shown for capacitor C_1. A higher value for C_1 would increase the low-frequency response. However, this would not be desired for a demodulator probe, since it is necessary to preserve the low-frequency modulation signal as an a-c voltage for measurement on the a-c circuit of the meter.

11. Once the probe has been made up, it should be checked for frequency response and demodulation capability as described in Chapter 4.

Voltage Dividers for Probes

It is possible to extend the voltage range of a probe when the probe must be operated with a-c voltages greater than the rating of the probe diode. In its simplest form, a probe voltage divider is a lower-value capacitor connected in series with the probe tip. This arrangement is similar to the capacitance-type voltage divider described in Sec. 2-4. The amount of voltage attenuation produced by the divider is determined by the capacitor value. For convenience in reading the meter scales, it is best to select a capacitor value that will provide a 10 : 1 voltage division. For a typical probe, the capacitance value should be in the range 50 to 150 pF.

One problem with any type of voltage divider is that the amount of attenuation is not consistent for all frequencies, particularly for probes used in low-frequency work (such as in audio-signal generators). This presents no difficulty if accurate readings are not required and the purpose of the divider is only to prevent damage. However, if accurate readings are necessary, it may be necessary to use more than one divider capacitor for various frequency ranges.

Another problem, especially with VOMs, is that the amount of attenuation will change each time the meter scales are changed. This is because the input circuit resistance of most VOMs will change with each scale change.

The problem is not so critical with electronic meters, since most electronic meters present a constant input resistance regardless of the meter scale.

The basic voltage-divider circuit is shown in Fig. 2-9. An improved voltage divider, especially useful for audio-frequency signals, is shown in Fig. 2-10. The circuit of Fig. 2-10 will provide a fairly constant amount of voltage division (or attenuation) over a band of frequencies. Such a voltage divider could be adjusted to provide a 10 : 1 attenuation across the audio-frequency range, say from 60 to 17,000 Hz. The following steps describe the calibration and fabrication procedure:

1. Connect the probe to a voltage source and to the meter. The voltage source should be at a frequency at which the divider is to be used.

2. Adjust the voltage source to some value that can be read conveniently on the meter (such as 10 V, 20 V, etc.).

3. Disconnect the probe from the voltage source. Connect the voltage-divider capacitor (Fig. 2-9) or capacitor-resistor combination (Fig. 2-10) to the probe tip. Connect the divider to the voltage source.

4. Check the amount of voltage attenuation produced by the divider. If the divider is used with a VOM and it is necessary to change scales to read the voltage attenuation, there may be an error in reading between scales. It is therefore recommended that attenuation be measured on one scale, if possible.

5. If the divider provides sufficient attenuation to prevent damage to the

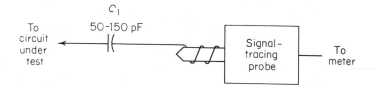

Fig. 2-9   Simple voltage-divider probe.

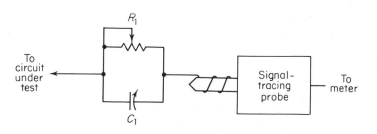

Fig. 2-10   Adjustable voltage-divider probe.

probe diodes and the amount of voltage division is exact (10 : 1, 5 : 1, etc.), then further calibration is unnecessary, and the divider can be put to use. However, if the amount of voltage division is not correct for any reason, the value of C_1 (or R_1 and C_1) must be changed.

6. A simple method is to use a variable capacitor and/or resistor and adjust for a desired attenuation. Then measure the value and substitute a fixed capacitor (or resistor) for the variable component.

7. In the capacitor-resistor divider of Fig. 2-10, the value of R_1 should be chosen to provide attenuation at the low end of the band, and C_1 should be chosen to provide the *same attenuation* at the high end of the band.

Amplifier for Probes

It is possible to increase the sensitivity of a probe with a transistor amplifier. Such an arrangement is particularly useful with a VOM for measuring small signal voltages. An amplifier is not usually required for an electronic meter, since such meters contain a built-in amplifier.

A practical transistor probe amplifier is shown in Fig. 2-11. This circuit will increase the sensitivity of the probe by at least 10 : 1 and should provide good response up to about 500 MHz. The circuit will not normally be calibrated to provide a specific voltage indication but is used to increase the sensitivity of the probe for signal-tracing purposes. The following steps describe fabrication procedures:

1. Connect the amplifier components to a signal-tracing probe (either half-wave or full-wave) and to the meter. Connect the transistor output circuit to the *current-input* terminals or leads of the VOM. Set the meter to measure current on its *lowest current range*. This would be on the order of 100 μA for a typical VOM.

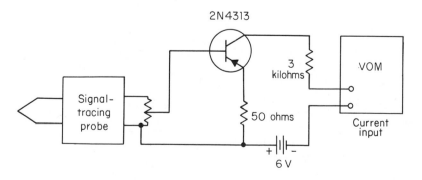

Fig. 2-11   Practical transistor probe circuit.

2. Connect the probe to a low-voltage source at the frequency at which
the probe and amplifier will be used. Do not apply a high voltage to
the probe. Because of the transistor amplification, a high voltage at
the probe will cause a large current to flow through the transistor
collector-emitter circuit and could possibly damage the VOM movement. With the VOM set to a 100-μA scale, the circuit of Fig. 2-11
should show readable indications with a 1000-μV (or less) signal applied to the probe.

3. Normally, there should be no current flow in the meter circuit when
the probe is not connected to a signal source. However, all transistors
have some leakage, and this may produce an indication on the meter
(especially when the meter is set to its lowest current scale). This can
be ignored for practical work. Also, it is possible to minimize the
leakage by selecting various transistors until a low-leakage transistor is
found.

4. The circuit of Fig. 2-11 will have little effect on the high-frequency
response of the probe. However, the low-frequency response should
be increased when the amplifier is added. This is because the amplifier
presents less of a current drain on the probe circuit than does the
meter input circuit.

3

Basic Operating Procedures and Techniques

The basic operating procedures for the various types of meters and probes described in Chapters 1 and 2 are discussed in this chapter. Because of the great variety of meters available, it is impossible to discuss the procedures for each make and model. Instead, typical units are described.

Later chapters describe how to use each type of meter to perform specific tests, alignments, and adjustments. There one finds instructions such as "set the meter to measure voltage" or "zero the ohmmeter circuit" or "read the output in decibels." It is assumed that the operator will become familiar with the particular equipment being used. This is *absolutely essential*, since the procedures given in this chapter can be used only as a general guide to operating meter controls and reading meter scales. The operator *must* understand each and every control and scale on the equipment in order to follow the instructions of later chapters. No amount of textbook instruction will make the operator an expert in operating meters; it takes actual practice.

NOTE

It is recommended that you establish a routine operating procedure or sequence of operation, for each meter in the shop or laboratory. This will save time and will familiarize you with the capabilities and limitations of your particular equipment, thus eliminating false conclusions based on unknown operating characteristics.

The first step in placing a meter in operation is reading the instruction manual for the particular meter. Although most instruction manuals are weak in application data, they do describe how to read meter scales, how to connect the test leads or probes, and the logical sequence for operating the controls. Therefore, the manual is of greater importance here, particularly if the operator is not familiar with the instrument's capabilities and limitations.

After the manual's instructions have been digested, they can be compared with the following procedures. Remember that the procedures set down in this chapter are general or typical and applicable regardless of the test to be performed or the type of meter used. On the other hand, instruction-manual procedures apply to a specific instrument. Therefore, if there is a conflict between the manual procedures and the following instructions, follow the manual. Remember the old electronics rule: when all else fails, follow instructions.

3-1 SAFETY PRECAUTIONS

Since a meter is an item of test equipment, certain precautions must be observed during its operation. Many of these precautions are the same as those to be observed for an oscilloscope or signal generator; others are unique to meters. Some of the precautions are designed to prevent damage to the meter or the circuit under test; others are to prevent injury to the operator. The following precautions are divided into two groups: general safety precautions and meter operating precautions. Both should be studied thoroughly and then compared to any specific precautions called for in the meter's instruction manual.

General Safety Precautions

1. Many electronic meters are housed in metal cases. These cases are connected to the ground of the internal circuit. For proper operation, the ground terminal of the meter should always be connected to the ground of the equipment under test. Make certain that the chassis of the equipment under test is not connected to either side of the a-c line (as is the case with some older a-c/d-c radio sets) or to any potential above ground. If there is any doubt, connect the equipment under test to the power line through an *isolation transformer.*

2. Remember, there is always danger inherent in testing electrical equipment that operates at hazardous voltages. Therefore, the operator should thoroughly familiarize himself or herself with the equipment under test before working on it, bearing in mind that high voltages may appear at unexpected points in defective equipment.

3. It is good practice to remove power before connecting test leads to high-voltage points. (High-voltage probes are often provided with alligator clips.) It is preferable to make all test connections with the power removed. If this is impractical, be especially careful to avoid accidental contact with equipment and other objects that can provide a ground. Working with one hand in your pocket and standing on a properly insulated floor lessens the danger of shock.

4. Filter capacitors may store a charge large enough to be hazardous. Therefore, discharge filter capacitors before attaching the test leads.

5. Remember that leads with broken insulation offer the additional hazard of high voltages appearing at *exposed* points along the leads. Check test leads for frayed or broken insulation before working with them.

6. To lessen the danger of accidental shock, disconnect test leads immediately after the test is completed.

7. Remember that the risk of severe shock is only one of the possible hazards. Even a minor shock can place the operator in danger of more serious risks, such as a bad fall or contact with a source of higher voltage.

8. The experienced operator continuously guards against injury and does not work on hazardous circuits unless another person is available to assist in case of accident.

Meter Operating Procedures

1. Even if you have had considerable experience with meters, always study the instruction manual of any meter with which you are not familiar.

2. *Never* measure a voltage with the meter set to measure current or resistance. To do so will burn out the meter movement. Likewise, *never* measure a current with the meter set to measure resistance.

3. Always start voltage and current measurements on the *highest* voltage or current scale. Then switch to a lower range as necessary to obtain a good center-scale reading.

4. Do not attempt to measure a-c voltages or currents with the meter set to measure dc. This could damage the meter movement and will pro-

duce errors in the meter readings. No damage will result (usually, but consult the instruction manual) if d-c voltages or currents are measured with the meter set to measure ac. However, the readings will be in error. The technique of measuring dc on the a-c ranges of a meter is discussed further in Chapter 4.

5. Use only shielded probes. Never allow your fingers to slip down to the metal probe tip when the probe is in contact with a "hot" circuit.

6. Avoid operating a meter in strong magnetic fields. Such fields can cause inaccuracy in the meter movement and could damage it. Most good-quality meters are well shielded against magnetic interference. However, the meter face is still exposed and is subject to the effects of magnetic fields.

7. Most meters have some maximum input voltage and current specified in the instruction manual. *Do not* exceed this maximum. Also, do not exceed the maximum line voltage or use a different power frequency on those meters which operate from line power.

8. Avoid vibration and mechanical shock. In the same manner as most electronic equipment, a meter is a delicate instrument.

9. Do not attempt repair of a meter unless you are a qualified instrument technician. If you must adjust any internal controls, follow the instruction manual.

10. Study the circuit under test *before* making any test connections. Try to match the capabilities of the meter to the circuit under test. For example, if the circuit under test has a range of measurements to be made (ac, dc, RF, modulated signals, pulses, or complex waves), it may be necessary to use more than one instrument. Most meters will measure dc and low-frequency ac. If an unmodulated RF carrier is to be measured, use an RF probe. If the carrier to be measured is modulated with low-frequency signals, a demodulator probe must be used. If pulses, square waves, or complex waves (combinations of ac, dc, and pulses) are to be measured, a peak-to-peak reading meter will provide the only *meaningful* indications.

11. Remember that all voltage measurements are made with the meter in *parallel* across the circuit (Fig. 3-17) and that all current measurements are made with the meter in *series* with the circuit (Fig. 3-18).

12. There are two standard *international operator warning symbols* found on some meters. One symbol, a *triangle with an exclamation point at the center*, advises the operator to refer to the operating manual before using a particular terminal or control. The other symbol, a *zigzag line simulating a lightning bolt*, warns the operator that there may be dangerously high voltage at a particular location, or that there is a voltage limitation to be considered when using a terminal or control. Examples of these symbols are shown in Fig. 3-6.

3-2 NONDIGITAL METER SCALES AND RANGES

Figures 3-1 and 3-2 show the scales and operating controls for a typical VOM (called a multitester) and a typical electronic voltohmmeter, respectively. The following notes describe each of the scales and provide information concerning their accuracy and use.

Ohmmeter Scales and Ranges

Note that the zero indication for the VOM ohmmeter scale is at the right and that on the electronic ohmmeter the zero indication is at the left. While this condition is typical, it will not be found on every make and model of meter.

Also note that the high-resistance or infinity end of the ohmmeter scale is cramped on both meters. Ohmmeter scales are always nonlinear. Therefore, ohmmeters provide their most accurate indications at midscale or near the low-resistance end.

Fig. 3-1   Radio Shack Multitester. (Courtesy Radio Shack.)

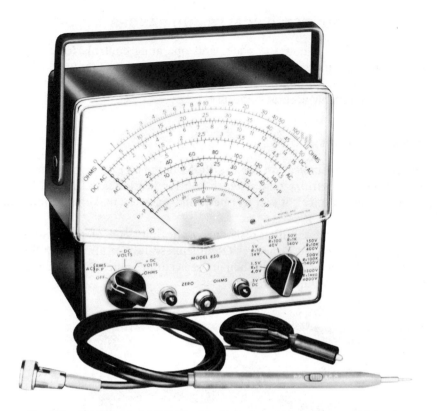

Fig. 3-2   Triplett Model 850 Electric Volt-ohmmeter. (Courtesy Triplett.)

In general, the ohmmeter scale is considered to be as accurate as the *d-c voltmeter* scale. However, in the case of a battery-operated VOM, the condition of the battery will affect accuracy. As a battery ages and its voltage output drops, the resistance indications will be *lower* than the actual value. For example, if a battery voltage drops to 90 percent of its *minimum* value, a 100-ohm resistor will produce a 90-ohm indication (approximately). This is true even though the ohmmeter is "zeroed" before making the measurement. Greatest accuracy will be obtained if the ohmmeter is zeroed on each range *just prior* to making the measurement.

In some meters, the ohmmeter scale is rated in *degrees of arc* rather than percentage of full scale as are the voltmeter and ammeter scales. For example, the d-c voltage accuracy of a meter could be ±3 percent of full scale. On the 100-V scale, this indicates an accuracy of ±3 V. As shown in Fig. 3-3, an indication of ±3 V (or a total of 6 V) corresponds to a certain number of degrees of arc. In turn, this arc defines the accuracy of the ohmmeter scale. Because the ohmmeter scale is nonlinear, it is possible that the

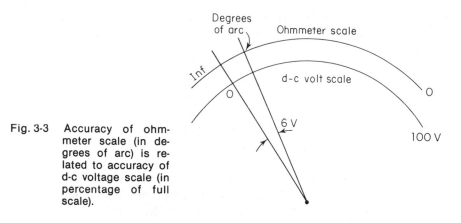

Fig. 3-3 Accuracy of ohm-meter scale (in degrees of arc) is related to accuracy of d-c voltage scale (in percentage of full scale).

error will not be constant over the entire scale. However, the error should not exceed the *rated accuracy* at any point on the scale.

D-C Scales and Ranges

The zero indication for the d-c scales is almost always on the left. Usually, the d-c scales are linear, with no cramping or bunching at either end. Note that there are three basic d-c scales (with maximums of 10, 50, and 250) on the VOM (Fig. 3-1). These same scales are also used for a-c measurements. Each of these scales serves multiple purposes, depending upon the position of the range selector. Therefore, the operator must make note of *both* the *scale reading* and the *range selector* position before the correct indication is obtained.

For example, the d-c readings of 150, 30, and 6 are all aligned. These readings are located on the 250, 50, and 10 scales, respectively. If the range selector is set to "0.5 DC.V," then the "30" reading applies, and the indicated value is 0.3 V. If the reading is the same, but with the range selector set to "2.5 DC.V," the "150" reading applies, and the indicated value is 1.5 V. Again, with the same reading, but with the range selector set to "10 DC.V," the "6" reading applies, and the indicated value is 6 V. Still using the same reading but with the range selector set to "50 DC.V," the "30" reading applies, and the indicated value is now 30 V. Finally, with the same reading but with the range selector set to "250 and UP DC.V," the "150" reading applies and the indicated value is 150 V.

The foregoing examples assume that the meter test leads are connected to the "-COM" and "+V.Ω.A" terminals. For voltages between 250 and 1000 V (the maximum limit of the meter) the "-COM" and "DC 1KV" terminals are used. Also, the examples above assume that the d-c voltage being measured is positive with respect to common or ground and that the meter function switch is accordingly set to "+DC.AC." If the d-c voltage to be measured is negative with respect to ground, the meter function switch must be set to "-DC."

The use of meter scales is often confusing to the inexperienced operator. Therefore, the meter scales should be *studied thoroughly* before attempting to use any meter for the first time.

Accuracy of the d-c scales is dependent upon the tolerance of the multiplier resistors, the accuracy of the movement, and the accuracy of the scales. In precision meters, the scales are matched to individual movements so that the combined movement and scale has a given accuracy. Usually, the multiplier (for voltage) and shunt resistors (for current) have an accuracy of ±1 percent or better. When this is coupled to a movement with a ±2 percent accuracy or better, the total accuracy is ±3 percent.

The accuracy of both d-c and a-c scales is usually specified as a *percentage of full scale* and not of the actual reading. This is another fact often overlooked by the inexperienced operator. For example, assume that a voltage is measured on the 50-V d-c scale (Fig. 3-1), that a reading of "10" is obtained, and that the rated accuracy is ±3 percent of full scale. The full-scale value is 50 V; therefore the absolute accuracy is ±1.5 V (50 × 3 percent). The 10-V reading could thus indicate an actual value of 8.5 to 11.5 V. Inexperienced operators often assume (incorrectly) that the 3 percent tolerance applies directly to the reading (or 0.3 V in the example given).

Several differences are noted between the d-c scales and ranges of the VOM (Fig. 3-1) and those of the electronic voltohmmeter (Fig. 3-2). First, the electronic meter is provided with a "zero" adjustment in addition to the "ohms" adjustment. Usually, this "zero" adjustment control must be set so that the movement's needle is aligned with zero each time the range selector is moved to another range position. There are several reasons for this. In vacuum-tube meters, there is a "contact potential" at the tube elements (particularly the grid) which changes with each position of the range selector. In transistor meters, there is a certain amount of leakage current which is subject to change. Also, there is a possibility of drift due to temperature change during warmup or over a long period of operation. All of these conditions are compensated for by the "zero" control (which should be set to provide meter zero each time the range selector is moved and thereafter periodically throughout operation).

Next, note that the meter function switch shows a "−d-c volts" position as well as a "+d-c volts" position. In a simple VOM, positive and negative voltages are adjusted for by reversing the leads. Usually, the black lead is connected to negative (−) and the red lead is connected to positive (+) when a positive d-c voltage is to be read. The leads are then reversed for a negative d-c voltage. (This should be verified, however, by consulting the instruction manual. If a manual is not available, check for correct lead connections with a battery or other d-c source where polarity is known. The VOM of Fig. 3-1 is provided with a polarity switch.)

In the electronic meter, the alligator clip is connected to ground, and the circuit voltage is measured with the probe (or prod) tip. The test leads are never reversed. Thus, it is necessary to have a negative and positive position for measurement of d-c voltages. On some electronic meters, the probe rather than the meter itself is provided with a switch for polarity reversing.

Next, note that the electronic meter does not have any current-measuring function. This is typical for most electronic meters. However, there are certain electronic meters that will measure current.

Finally, note that the electronic meter has a special *zero-center* scale (bottom scale). This scale permits both positive and negative voltage indications to be displayed on either side of a zero center without reversing the function switch or the test leads. In use, the function switch is set to " + d-c volts," and the "zero" knob is adjusted until the meter needle is aligned with the zero center (with no voltage applied). A positive voltage will deflect the needle to the right (+), and a negative voltage will move the needle to the left (−). The scale divisions are arbitrary and have no relation to actual voltage. However, they do provide a relative measure of voltage on either side of zero. The zero-center scale permits the meter to be used as a null meter in a bridge circuit, as an FM-receiver detector-alignment tool, as well as for other functions described in later chapters.

A-C Scales and Ranges

The zero indication for the a-c scales is almost always on the left. In some older models, the a-c scales are somewhat nonlinear, with some cramping or bunching on the low end (or least full-scale value). This is because the rectifier circuits required for a-c measurement are nonlinear. Both half-wave and full-wave rectifier circuits are nonlinear. Nonlinearity will be more pronounced when the multiplier resistance is small (low-voltage ranges) than when the multiplier resistance is large (high-voltage ranges). This condition is known as *swamping* effect.

The a-c scales of a meter are never more accurate and usually less accurate than the d-c scales. This is because the inaccuracy of the a-c rectifier circuit must be added to the inaccuracy of the d-c circuit (multiplier, movement, and scales). Accuracies for a typical meter are ±3 percent of full scale for dc and ±5 percent of full scale for ac. An electronic meter will sometimes have the same accuracy for both a-c and d-c scales (typically ±3 percent).

The effects of *frequency* must be considered in determining the accuracy of a-c scales. A typical VOM will provide accurate a-c voltage indications from 15 Hz up to 10 kHz, possibly up to 15 or 20 kHz, but rarely beyond that frequency. This means that VOM readings in the high audio range may be inaccurate. It should be noted that an a-c meter may provide readings well beyond its maximum rated frequency, but these readings will not necessarily be accurate. A typical electronic meter will provide accurate a-c voltage indications from 15 Hz up to about 3 MHz. Of course, the frequency range of a meter can be extended by use of an RF probe. However, the probe and meter must be calibrated together as described in Chapter 2.

As shown in Fig. 3-2, many electronic meters permit a-c voltages to be read as an RMS value or a peak-to-peak value, whichever is convenient. Note that the meter circuit responds to the average value in either case but that the scales provide for RMS and peak-to-peak indications.

Note also that the RMS scales are accurate *only for a pure sine wave*. If there is any distortion or if the voltage contains any component other than a pure sine wave, the readings will be in error. On the other hand, peak-to-peak readings will be accurate on *any type of waveform*, including sine waves.

A-c voltage measurements are usually made with a *blocking capacitor* in series with one of the test leads. On some VOMs, the test lead is connected to a terminal marked "output" or some similar function. This blocks any dc present in the circuit being measured from passing to the meter circuit. Such dc may or may not damage the meter, depending upon conditions.

A-c voltage can also be measured on most meters without the blocking capacitor. On a typical VOM, this is accomplished by connecting the test leads to the same terminals as for d-c voltage measurements (usually marked " + " and " − " or "common" and " + ").

Some older meters have half-wave rectifier circuits or use a half-wave probe circuit. On such meters, the a-c voltage readings may be affected by a condition known as *turnover*. This occurs when there are *even harmonics* present in the voltage being measured. Turnover will show up as different a-c voltage readings when the test leads are reversed. Turnover should not occur when the harmonics are odd, when there are no harmonics, or when a full-wave rectifier is used.

Decibel Scales and Ranges

Most VOMs are provided with decibel (dB) scales. Actually, the a-c voltage circuit is used in the normal manner, except that the readout is made on the dB scales. Inexperienced operators are often confused by the dB scales. The following notes should clarify their use.

The dB scales represent *power ratios* and not voltage ratios. In most cases, 0 dB is considered as the power of 1 mW (0.001 W) across a 600-ohm pure-resistive load. This also represents 0.775 V RMS across a 600-ohm pure-resistive load. The term *decibel meter*, or dBM, is sometimes used to indicate this system (1 mW across 600 ohms).

The dB scale is related directly to one of the a-c scales, usually the lowest scale. The VOM range selector must be set to that a-c scale if readings are to be taken directly from the dB scale. If another a-c scale is selected by the range selector, a certain dB value must be added to the indicated dB-scale value. For example, in the VOM of Fig. 3-1, the dB scale is related directly to the 5 V a-c scale. If the range selector is set to 5 V a-c, the dB scale may be read out directly. Note that 0 dB is aligned with the 0.775-V point on the 5 V a-c scale. If the range selector is set to 10, 50, 250, or 1000 V a-c, it is necessary to add 6, 20, 34, or 46 dB to the indicated dB-scale reading. These values are printed on the meter face (lower right-hand corner) and are applicable to that meter only. Always consult the meter face (or instruction manual) for data regarding the dB scales.

Note that dB scale readings *will not be accurate* if (1) the voltages are other than pure sine waves, (2) the load impedances are other than pure-resistive, and (3) the load is other than 600 ohms.

If the load is other than 600 ohms, it is possible to apply a correction factor. The decibel is based on this mathematical function:

$$dB = 10 \log \frac{\text{power output}}{\text{power input}}$$

Since the power will change by the corresponding ratio when resistance is changed (power will increase if resistance decreases and voltage remains the same), it is possible to convert the function to $10 \log R_2/R_1$, where R_2 is 600 ohms and R_1 is the resistance value of the load.

For example, assume that the load resistance is 500 ohms instead of 600 ohms, and a 0-dB indication is obtained (0.775 V RMS).

$$\frac{600}{500} = 1.2$$

$$\log 1.2 = 0.0792$$

$$10 \times 0.0792 = 0.792$$

Therefore, 0.792 (or 0.8 for practical purposes) must be added to the 0-dB value to give a true reading of 0.8 dB.

Table 3-1 lists correction factors to be applied for some common load impedance values. This table shows the amount of dB correction to be added to the indicated dB value when the load impedance is other than 600 ohms. For example, if the load impedance is 300 ohms, a $+3$ dB must be added to the indicated value. This can be verified using the previous equation:

Resistive Load at 1 kHz	dBM
500	+0.8
300	+3.0
250	+3.8
150	+6.0
50	+10.8
15	+16.0
8	+18.8
3.2	+22.7

$$\frac{600}{300} = 2$$

$$\log 2 = 0.3010$$

$$10 \times 0.3010 = 3.01$$

Therefore, 3.01 (or 3.0 for practical purposes) must be added to the indicated dB value to give a true reading.

Figure 3-4 shows the relationship between a-c voltages (RMS), decibels, and power (in milliwatts across a 600-ohm pure-resistive load). This illustration can be used when a particular meter does not have a dB scale but does show RMS voltages. Of course, the correction factor of Table 3-1 must be applied to the value if the load is not 600 ohms.

For example, assume that a 2.5-V RMS signal were measured across a 50-ohm load. Figure 3-4 shows that 2.5 V RMS equals + 10 dB (actually slightly higher). Table 3-1 shows that a + 10.8-dB correction factor must be added, resulting in a 20.8-dB, or possibly a 21-dB true reading.

Note that if the load resistance is greater than 600 ohms, the power will be reduced, and the correction factor must be *subtracted*. Note also that dBM or *decibel-meter* values are usually based on a frequency of 1 kHz, since this is the frequency where the decibel system most nearly corresponds to the characteristics of the human ear.

There is often confusion in making dB measurements at the input and output of a particular circuit (such as an amplifier) to find gain, particularly with load impedances. The following rules should clarify this problem.

If the input and output impedances are 600 ohms (or whatever value is used on the meter scale), no problem should be found. Simply make a dB reading at the input and at the output (under identical conditions), subtract the smaller dB reading from the larger, and note the dB gain (or loss). For

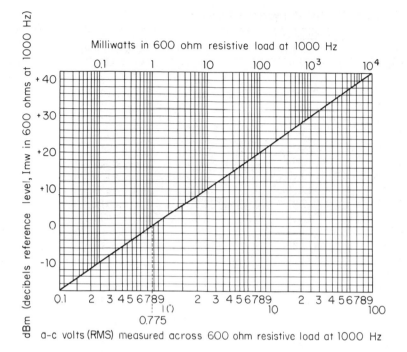

Fig. 3-4   Graph for conversion of RMS voltage to dBm values. (Courtesy RCA.)

example, assume that the input shows 3 dB, with 13 dB at the output. This represents a 10-dB gain. If the output had been 3 dB, with a 13-dB input, there would have been a 10-dB power loss.

The dB gain (or loss) can be converted into a power ratio (or voltage or current ratio) by means of Table 3-2. Using the example of a 10-dB gain (or loss), this represents a power ratio of 10 and a voltage or current ratio of 3.1623.

If the input and output load impedances are not 600 ohms but are equal to each other, the relative dB gain or loss is correct, even though the absolute dB reading is incorrect. For example, assume that the input and output load impedances are 50 ohms and that the input shows 3 dB with 13 dB at the output. Table 3-1 shows that 10.8 dB must be added to the input and output readings to obtain the correct dBM absolute value. However, there is still a 10-dB difference between the two readings. Therefore, the circuit shows a 10-dB gain, and the power (or voltage/current) ratios of Table 3-2 still hold.

If the input and output load impedances are not equal, the relative dB gain or loss indicated by the meter scales will be incorrect. For example,

TABLE 3-2 Decibel Conversion Chart

Power Ratio	Voltage and Current Ratio	Decibels (+) (−)	Voltage and Current Ratio	Power Ratio
1.000	1.000	0.0	1.000	1.000
1.023	1.012	0.1	.9886	.9772
1.047	1.023	0.2	.9772	.9550
1.072	1.035	0.3	.9661	.9333
1.096	1.047	0.4	.9550	.9120
1.122	1.059	0.5	.9441	.8913
1.148	1.072	0.6	.9333	.8710
1.175	1.084	0.7	.9226	.8511
1.202	1.096	0.8	.9120	.8318
1.230	1.109	0.9	.9016	.8128
1.259	1.122	1.0	.8913	.7943
1.585	1.259	2.0	.7943	.6310
1.995	1.413	3.0	.7079	.5012
2.512	1.585	4.0	.6310	.3981
3.162	1.778	5.0	.5623	.3162
3.981	1.995	6.0	.5012	.2512
5.012	2.239	7.0	.4467	.1995
6.310	2.512	8.0	.3981	.1585
7.943	2.818	9.0	.3548	.1259
10.00	3.162	10.0	.3162	.10000
12.59	3.548	11.0	.2818	.07943
15.85	3.981	12.0	.2515	.06310
19.95	4.467	13.0	.2293	.05012
25.12	5.012	14.0	.1995	.03981
31.62	5.632	15.0	.1778	.03162
39.81	6.310	16.0	.1585	.02512
50.12	7.079	17.0	.1413	.01995
63.10	7.943	18.0	.1259	.01585
79.43	8.913	19.0	.1122	.01259
100.00	10.000	20.0	.1000	.01000
10^3	31.62	30.0	.03162	.00100
10^4	10^2	40.0	10^{-2}	10^{-4}
10^5	316.23	50.0	3.162×10^{-3}	10^{-5}
10^6	10^3	60.0	10^{-3}	10^{-6}
10^7	3.162×10^3	70.0	3.162×10^{-4}	10^{-7}
10^8	10^4	80.0	10^{-4}	10^{-8}
10^9	3.162×10^4	90.0	3.162×10^{-5}	10^{-9}
10^{10}	10^5	100.0	10^{-5}	10^{-10}

assume that the input impedance is 300 ohms, the output impedance is 8 ohms, the input shows + 7 dB, and the output shows + 3 dB, on the scales of a meter using a 600-ohm reference.

There is an apparent loss of 4 dB (7 dB input − 3 dB output). However, by referring to Table 3-1, it will be seen that the 300-ohm input (7 dB) requires a correction of + 3 dB (giving a corrected input of + 10 dB), and the 8-ohm output (3 dB) requires a correction of + 18.8 dB (giving a corrected output of + 21.8 dB). Thus, there is an actual gain of + 11.8 dB.

3-3 DIGITAL METER DISPLAYS AND RANGES

Figures 3-5 and 3-6 show the displays and operating controls for a portable digital multitester (called a digital voltmeter) and a bench-type digital multitester, respectively. The following notes describe each of the displays and control, and provide information concerning their accuracy and use.

Digital Displays or Readouts

The first major difference between digital meters (Figs. 3-5 and 3-6) and nondigital meters (Figs. 3-1 and 3-2) is that all digital meter readings or values are shown on the digital displays. Thus, the same digital displays

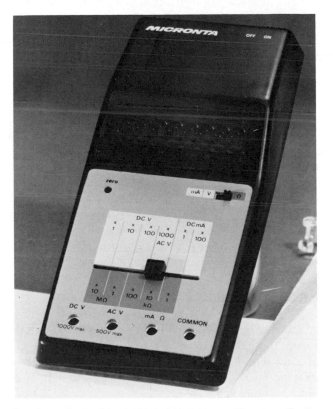

Fig. 3-5   Radio Shack Digital Volt Meter. (Courtesy Radio Shack.)

Fig. 3-6   Heathkit Auto Ranging Digital Multimeter. (Courtesy Heath Company).

show voltage, current, and resistance, depending upon the settings of con-trols.

Nonautoranging Digital Meter

The portable meter of Fig. 3-5 is nonautoranging. That is, the range must be set by means of a range selector. The display has a *fixed decimal point*, and it is necessary to multiply the reading by the number next to the range selec-tor (a horizontal slide switch). For example, to measure d-c voltage, the function switch must be set to mA/V; then a d-c volts range of ×1, ×10, ×100, or ×1000 is selected. The reading shown on the display must be multiplied by the range selector number. That is, if the display is 0.987 and the d-c volts range switch is set to ×10, the true indication is 9.87 V.

Voltage and Current Overload

The portable meter (Fig. 3-5) is provided with an overload indication. With all ranges (except the 1000 V d-c and a-c volts ranges) it is possible to use up to ±1999 of a displayed value. When ±1999 is exceeded, the display will show a = 000 or ≡ 000 indication. This means that it is necessary to manually select a higher range.

Test Leads and Polarity

When the 1000 V d-c range is selected, the display reads up to ± 1000. When a-c volts is selected, the display reads up to 500. Note that the same position of the range selector is used for a-c volts measurements, and for d-c volts × 1000. Thus, the choice between a-c and d-c measurements must be made by means of the test leads. Normally, the black test lead is connected to the COMMON jack, and the red test lead is connected to d-c volts or a-c volts, as required.

When the test leads are connected into a circuit to measure d-c voltages, a (−) or minus polarity sign appears in the display if the voltage being measured is negative with respect to the COMMON lead. If no polarity sign appears, the voltage being measured is positive with respect to the COMMON lead.

A-C Voltage Measurements

To measure a-c voltage on the portable meter of Fig. 3-5, the function switch is set to mA/V, the range switch is set to a-c volts (which is also d-c volts × 1000), the black test lead is connected to COMMON, and the red lead is connected to a-c volts. Under these conditions, the meter senses the *mean value* of the voltage being measured and is calibrated to read the RMS value of a sine wave (refer to Sec. 1-4). Any direct current in the circuit being tested will be added to the alternating current and will affect the reading. This problem can be overcome by first measuring the circuit for any d-c voltage and then subtracting the d-c voltage from the a-c reading. An alternative procedure is to connect a capacitor in series with the red test lead. This blocks any dc from the meter circuits and leaves only the a-c voltage.

Current Measurements

To measure direct current on the portable meter of Fig. 3-5, the function switch is set to mA/V, the range switch is set to d-c mA (either 1 or 100), the black test lead is connected to COMMON, and the red lead is connected to mAΩ . As discussed in Sec. 3-10, when measuring current, you must break the circuit and connect the test leads to the two circuit connection points. As in the case of voltage, a (−) or minus polarity sign appears in the display if the current being measured is negative with respect to the COMMON lead. Note that the meter of Fig. 3-5 cannot be used to measure alternating current directly.

Four extra ranges of d-c measurement are available by using the mA Ω socket and selecting one of the d-c volts ranges. The current range value is obtained by multiplying the d-c volts range multiplier by 0.1 μA. These very low current ranges are extremely useful for measuring leakage in solid-state devices since currents down to 100 pA can be resolved on the digital readout.

Resistance Measurements

To measure resistance on the portable meter of Fig. 3-5, the function switch is set to Ω, the range switch is set to one of the two MΩ positions (1 or 10) or to one of the three kΩ positions (1, 10, or 100), the black test lead is connected to common, and the red lead is connected to mAΩ. Always make certain that all power is removed from the circuits before making any resistance measurements, as discussed in Sec. 3-8.

Note that the resistance measuring circuit of the meter applies a known value of constant current through the unknown resistance, and then measures the voltage across the circuit (as discussed in Sec. 1-12). Because of this, the resistance ranges can be used to measure *forward voltage drop of semiconductors* (refer to Sec. 6-1). When the meter is set to measure resistance and the test leads are connected across a semiconductor diode so that the diode is forward-biased, the reading on the display is the forward voltage drop, expressed in volts. For example, when the range selector is set to k$\Omega \times$ 1, the constant current value is 1 mA. If the reading is 0.5 (indicating a diode forward resistance of 500 ohms), the forward voltage drop is 0.5 V.

The resistance ranges of the portable meter are also provided with an overload indication. The meter will show flashing bars until a resistance lower than the maximum reading of the range is connected. Both *positive overload* and *negative overload* can occur. A positive overload is where the resistance value is greater than the maximum reading of the range. A negative overload is where the resistance value is lower than the resolution provided by the last digit of the scale. For example, on the k$\Omega \times$ 10 resistance range, the extreme right-hand digit equals 10 ohms. If a 5-ohm resistor is connected on the scale, there will be a negative-overload indication.

A positive overload always produces flashing bars. A small negative overload also gives flashing bars, but a large negative overload can produce fixed bars. Should these fixed bars appear on the resistance function, indicating a large negative overload, the condition can be corrected by temporarily placing the test leads across a large resistance.

Zero Adjustment

Many digital meters are provided with a zero adjustment, such as the front-panel ZERO control shown in Fig. 3-5. This control is accessible with a small screwdriver, and does not normally require repeated adjustment. The zero setting of the meter is checked by setting the range switch to × 1000, and the function switch to mA/V. With no input and no test leads connected, the display should read 000, or possibly − 000. If not, the ZERO control is adjusted until the display is 000 or − 000. A slight jitter to a reading of 001 may occur. This is normal and will not affect meter accuracy.

Accuracy and Resolution

The accuracy of a digital meter is closely related to resolution. The accuracy figures or specifications for digital meters (and most other digital readout instruments) as a percentage, ± 1 count (or possibly ± 2 counts for a-c measurements). Thus, it is necessary to determine what each count represents on each scale to determine accuracy. For example, in the meter of Fig. 3-5, the accuracy of the 1-V range for d-c voltage measurements is rated as 1 percent ± 1 count. (In this case, the 1 percent applies to the reading.) The resolution for the 1-V d-c range is 1 mV (the extreme right-hand digit equals 1 mV). As a result, if the display is 0.987, the true voltage is within 1 percent of the reading (within 9.87 mV) ± 1 mV. This produces an area of uncertainty of about 22 mV (almost 11 mV above or below the display reading).

Autoranging Digital Meter

The meter of Fig. 3-6 is capable of full autoranging. That is, the range changes automatically to suit the value being measured, once a particular function (voltage, current, resistance) has been selected by a FUNCTION switch. The display has a *floating decimal point* and is therefore *direct reading*. It is not necessary to multiply the reading by a scale factor. For example, to measure d-c voltage, the FUNCTION switch must be set to volts dc, the test leads connected to the COM and IN jacks, the RANGE switch set to AUTO, and then the test lead tips connected to the circuit. The digital display will move automatically to the appropriate one of five d-c voltage ranges (± .2, ± 2, ± 20, ± 200, ± 2000). That is, if the display is + 9.870, the true indication is + 9.87 V (the display is in the ± 20 range).

Maximum Displays

Each digital range is four-place. Thus, the maximum display for the .2 range is .1999, the 2 range is 1.999, the 20 range is 19.99, and the 200 range is 199.9. The maximum display for the 2000-V range is 1000, since this is the maximum voltage capability of the meter.

Range Override

It is possible to override the autoranging function. When the RANGE switch is set to HOLD, the display will remain in the last range selected, or move to a lower range. Thus, if the RANGE switch is moved to HOLD when a reading of +9.870 is obtained in AUTO, the display will remain in the ±20 range (or drop to a lower range, if necessary). This feature is very useful when making repetitive voltage measurements in a particular range. When the meter is set to AUTO, the display will go through each range (as necessary) to reach the appropriate range for each measurement. This can be a waste of time if all the measurements are in the same range. However, the *total response time* of the display, including full autoranging, is less than 3 sec.

Minimum Displays

Most digital meters have some minimum display limits. For example, in the ±2 range, the minimum display limit is 0.180 V. Voltages below this value will not appear on the ±2 range. However, this presents no problem to the operator, since in either HOLD or AUTO mode, the display will automatically drop down to the next lower range (±.2) and produce the appropriate reading.

Overload and Overrange Indications

Where there is an overload or overrange condition, all display elements (except the kΩ and MΩ indicators) will flash repeatedly. Thus, when measuring voltages greater than 1000 (an overload) in the AUTO mode, the display elements continue to flash until the voltage is removed. The display elements also flash to indicate an overrange condition when the RANGE switch is in HOLD and the selected range is exceeded. The flashing numbers displayed during the overload/overrange conditions may be any number from 000 to 1000. The numbers are essentially meaningless and are not to be considered an accurate indication of value.

Test Leads and Connections

In normal operation, the red test lead is connected to the IN jack, and the black test lead is connected to the COM jack. Note the standard international operator warning symbols (discussed in Sec. 3-1) located next to the jacks. These symbols indicate that no more than 1000 V be connected to the jacks and that the manual should be consulted before connecting any leads to the jacks.

Although the meter is supplied with test leads, a number of alternative test leads can be used. The manual recommends "twisted pair," coaxial, or shielded cable test leads when measurements must be made in the presence of strong RF signals. This subject is discussed further in Sec. 3-7.

The meter can also be used with an RF probe (Sec. 2-5) or with high-voltage probes (Secs. 2-3 and 2-4). When a high-voltage probe is used, make certain to multiply the indicated reading by the probe division factor.

Polarity Indications

In addition to the four-place readout, the display is provided with polarity indications or signs. When measuring d-c voltage or current (FUNCTION switch set to volts dc or mA dc), a (+) or positive polarity sign appears in the display if the voltage or current being measured is positive with respect to the COM (black) lead. A (−) or negative polarity sign appears for voltages or currents that are negative with respect to the COM lead. No polarity indications or signs are displayed for a-c and resistance measurements.

Zero Adjustment

For the most accurate results, the meter should be zeroed at regular intervals, particularly when the function is changed. For d-c voltage and current functions, the meter is zeroed by connecting the test leads, and (if necessary) adjusting the front-panel ZERO control to display + .0000. An occasional display of + .0001 or − .0001 is normal. The resistance function is zeroed in essentially the same way. However, it is possible for the test leads to show some small resistance, typically 0.1 or 0.2 ohm (which appears as .0001 or .0002 on the k Ω range). The zero control is inoperative on the a-c voltage and current functions.

D-C Voltage Measurement Considerations

The following considerations apply to the autoranging multimeter of Fig. 3-6. However, similar considerations apply to a variety of similar meters when measuring d-c voltage.

The accuracy specification for d-c voltage is ±(0.05 percent of reading + 0.10 percent of range + 1 count). For example, a display reading of 1.000 V dc from a low-impedance source will have an uncertainty of ±0.0035 V dc.

The input resistance of the meter on all volts d-c ranges is 10 megohms. Measurements of relatively high resistances could cause a significant reading error. The amount of error due to meter loading can be determined by the equation

$$\text{percent error} = - \frac{R_S}{R_S + 10 \text{ megohms}} \times 100$$

where R_S is the resistance being measured (called the source resistance).

For example, a source resistance (R_S) of 10 kilohms results in a loading error (reading error due to the meter load) of −0.1 percent. The error has a "−" or minus sign, since the loading reduces the voltage under "load" from the "unloaded" (meter not connected) value. The loading error becomes very significant for source resistances above 100 kilohms.

Over an extended period of operating time, there may be some variation in the shorted input zeroing on the .2 V dc range to which the meter automatically ranges when the test leads are shorted together. The least-significant (right-hand) digit may, because of ambient temperature changes, vary positive or negative from +.0000. The right-hand digit is equivalent to 100 μV as the resolution of the .2 V d-c range. Since voltage measurements on higher ranges are more common, the effect of this variation is reduced, in decade steps, to the extent that even several hundred microvolts are an insignificant part of values usually measured. Thus, it is typically not an important effect to measurement accuracy when this occurs. For critical measurements on the .2 V d-c range, a "touch-up" of the front panel ZERO control will remove this potential error.

The ZERO control may be used as an *offset adjustment* to remove small, residual voltages when making differential or null measurements. However, the range of adjustment decreases in decade steps such that only a very limited "zeroing" capability is provided on higher ranges. Therefore, zero offset is recommended only for the lowest (.2 and 2 V d-c) ranges. Readjust the ZERO control when zero offset measurements are complete.

When the meter inputs are open-circuited on the .2 V d-c range, there will be several counts displayed, because of bias currents in the measuring circuitry. This is normal and will not produce a significant measurement error when the leads are connected to a low resistance (less than 1 megohm).

A-C Voltage Measurement Considerations

The following considerations apply to the autoranging multimeter of Fig. 3-6. However, similar considerations apply to a variety of similar meters when measuring a-c voltages.

The maximum a-c voltage allowable between the IN and COM jacks is 700 V RMS. This corresponds to about 1000 V peak. When measuring a-c voltage, any input other than a pure sine wave will cause an error because the a-c converter is average-sensing and RMS-calibrated. Square waves, sawtooth waves, and so on, can best be measured with an oscilloscope, as discussed in Sec. 5-7. When the volts a-c function is selected, the meter is capacitively coupled to the circuit so that any dc is blocked from entering the meter. Thus, only the a-c voltage present in the circuit appears on the meter.

The accuracy specification for volts ac is $\pm$ (0.25 percent of reading + 0.15 percent of range + 1 count) over the frequency range applicable to the volts a-c range being used. For example, a display reading of 1.000 V ac from a low-impedance source will have an uncertainty of ± 0.0065 V ac over the frequency range 40 Hz to 20 kHz.

The input impedance of the meter on all volts a-c ranges is *frequency dependent* and can be represented as 10 megohms, in parallel with approximately 90 pF. This corresponds to approximately 9.47 megohms at 60 Hz. Measurements at relatively high source resistances can cause a significant reading error. The meter input impedance at other frequencies may be determined by the following expression:

$$Z_{IN} = \frac{10 \text{ megohms}}{\sqrt{1 + (5.655 \times f)^2}}$$

where Z_{IN} is the effective meter input impedance and f the frequency in kHz.

The loading error caused by the impedance of the circuit or device being measured (called the source impedance) can be determined by the following expression:

$$\text{percent error} = -\left(\frac{Z_{source}}{Z_{source} + Z_{IN}} \times 100 \right)$$

The loading error at low frequencies (below 100 Hz) can be very significant for source impedances above 100 kilohms, and at higher frequencies (in the order of 10 kHz) for source impedances even as low as 2000 ohms. The error will always have a " $-$ " sign, since the loading will always reduce the voltage under load from its unloaded value.

When the meter inputs are open-circuited on the .2- or 2-V a-c ranges, there will be a significant number of counts displayed, because of stray line voltage radiation. This is normal and will not produce a measurement error when the leads are connected to a low-impedance source.

In the HOLD mode (with the meter input shorted), the display may show a 1- or 2-count "residual" on the higher ranges of the VAC function. The accuracy specification for this function includes the 1- or 2-count offset.

Direct-Current Measurements

The following considerations apply to the autoranging multimeter of Fig. 3-6. However, similar considerations apply to a variety of similar meters when measuring direct current.

The meter is fuse-protected for a 3-A maximum direct current on the mA d-c ranges. If this is exceeded, the fuse will open, the display will read zero, and the circuit under test will be opened.

The accuracy specification for mA dc is $\pm$ (0.15 percent of reading + 0.10 percent of range + 1 count). For example, a display reading of 1.000 mA dc will have an uncertainty of ± 0.0045 mA dc.

When current is measured, the multimeter will, to some degree, affect operation of the circuit being tested. This effect, known as "insertion loss," causes a voltage drop and will reduce the actual circuit current to the current displayed on the meter. This error must be considered if the resistance of the circuit under test is not at least 1000 times the shunt resistor for the range being used. For example, on the .2 mA d-c range, the shunt resistance is 1000 ohms. Therefore, a source resistance of 100 kilohms results in an insertion loss error of approximately 1 percent of the reading. Insertion loss error for other source resistances can be determined by the following relationship:

$$\text{percent error} = -\left(\frac{R_{\text{shunt}} + 0.15 \text{ ohm}}{R_{\text{source}} + R_{\text{shunt}} + 0.15 \text{ ohm}} \times 100 \right)$$

where 0.15 ohm is the approximate fuse and wiring resistance of the meter and R_{shunt} = 1000, 100, 10, 1, and .1 ohm for the .2, 2, 20, 200, and 2000 mA d-c ranges, respectively.

The insertion loss error will always have a " – " sign, since the "inserted" current is always less than the "not inserted" current. To reduce this effect, use the highest range possible consistent with the measurement resolution required (by means of the HOLD mode).

The ZERO control may be used as an offset adjustment to remove small, residual currents when making differential or null measurements.

However, since this adjustment has the same (count) effect on all ranges, the offset technique is applicable only on the range where the initial offset adjustment is made. Readjust the ZERO control (on the volts d-c mode) when these types of measurements are completed.

Over an extended period of operating time, there may be some variation in the zeroing of the .2 mA d-c range to which the meter automatically ranges when the test leads are open-circuited. The least-significant (right-hand) digit may, because of ambient temperature changes, vary positive or negative from a display of + .0000. Periodically check the zeroing (on the volts d-c mode) and "touch up" if necessary.

When you attempt to measure direct current with a substantial a-c or pulse component superimposed on the direct current, a significant error will result if the peak-to-peak variation in current exceeds three times full-scale direct current of the range being used (1.5 times on the 2000 mA d-c range). A higher range, using the HOLD mode, will minimize this error.

Alternating-Current Measurements

The following considerations apply to the autoranging multimeter of Fig. 3-6. However, similar considerations apply to a variety of similar meters when measuring alternating current.

The meter is fuse-protected for a 3-A maximum alternating current on the mA AC ranges. If this is exceeded, the fuse will open, the display will read zero, and the circuit under test will be opened. When measuring alternating current, any input other than a pure sine wave will cause an error because the a-c converter is average-sensing and RMS (sine wave)-calibrated. Square waves, sawtooth waves, and so on, can be measured best with an oscilloscope, as discussed in Sec. 5-7. When the mA a-c function is selected, the meter is capacitively coupled to the circuit so that any dc is blocked from entering the meter. Thus, only the ac present in the circuit appears on the meter.

The accuracy specification for mA ac is ±(0.30 percent of reading + 0.10 percent of range + 2 counts) over the frequency range 40 Hz to 10 kHz (5 kHz on 2000 mA a-c range). For example, a display reading of 1.000 mA ac will have an uncertainty of ±0.008 mA ac.

As in the case of direct current, the meter will cause some insertion loss when measuring alternating current. The effects of a-c insertion loss can be calculated in exactly the same way as for d-c insertion loss.

The mA a-c ranges are particularly affected by high *common-mode voltage*, which is produced when making current measurements in circuits not directly at earth (power line) levels. This problem is discussed further next.

Ground Loops, Floating Measurements, and Common-Mode Problems

One of the problems in any meter, particularly differential and digital meters that operate from a power line, is an effect known as *common-mode insertion*. This is essentially a form of noise or undesired signals appearing at the input of an instrument, due to circulating *ground currents* between the meter and circuit under test.

One of the major causes of common-mode signals is induced ground currents, usually at the a-c power-line frequency (typically 60 Hz). These signals can generate a potential of several volts between the circuit ground and the meter chassis or case ground. Unless bypassed, these currents will cause a voltage to appear at the input. This voltage could be larger than the voltage (or current) being measured. In any event, the ground loop voltage (or current) will be added to the measured voltage, resulting in an improper reading.

A typical common-mode problem is shown in Fig. 3-7. Here, a bridge-type transducer is connected to the input of a digital meter. The output voltage across R_H and R_L of the bridge is applied to the measuring circuit input and results in the desired reading. Note that there is another circuit (or

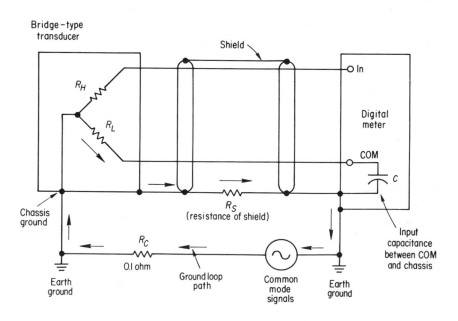

Fig. 3-7 Typical common mode signal problem.

a-c path) through R_L, C, and R_C. Capacitor C is the input capacitance between the COM terminal of the meter and the chassis or case ground. Resistor R_C is the resistance between the earth grounds of the transducer and the meter. In the circuit of Fig. 3-7, the junction R_H and R_L (the circuit being measured) is shown connected directly to the chassis or case ground of the transducer. This connection need not be direct, but can be reactive (such as the capacitance or inductance between the junction and case or chassis). When there is no direct connection to ground, the circuit is said to be *floating*, and the measurement is a *floating measurement*.

Although the values of C and R_C are low, they do exist and provide an a-c path for any signals or currents present in the ground line (known as the *ground loop*). These currents or signals cause an undesired voltage to be developed across R_L. This voltage is mixed with the desired transducer output and applied to the measuring circuit of the meter. Such voltages or signals are a source of error. The error is most noticed when measuring current (particularly alternating current) but is also a problem on the voltage ranges.

The capacity of a meter to reject a common-mode signal and thus reduce the undesired signal currents is called *common-mode rejection* or *common-mode rejection ratio* (CMR or CMRR). Note that the term "CMRR" is usually applied to the rejection capability when both inputs are floating. Neutral-mode rejection ratio or NMRR is often applied to indicate the rejection capability when one input is grounded. Usually, CMRR and NMRR are specified in decibels at some frequency or range of frequencies. For example, a typical common-mode rejection figure is 130 dB minimum at 60 Hz. This means that the undesired signal effect of a given common-mode signal is reduced by 130 dB or more. Thus, the effect of a 100-V, 60-Hz common-mode signal is reduced to that produced by a 33-μV maximum equivalent signal. As a point of reference, the meter of Fig. 3-6 has an NMRR rating of greater than 80 dB (over a frequency range of 50 Hz to 20 kHz) a CMRR rating of greater than 100 dB at dc, and greater than 70 dB (over a frequency range of 50 Hz to 1 kHz). These specifications apply to the d-c voltmeter function when the meter is operated on line power. The ratings improve to greater than 120 dB at dc and greater than 100 dB (over a frequency range of 50 to 20 kHz) when the meter is operated on battery power (and the power cord is disconnected).

From a study of the foregoing specifications, it can be seen that battery operation (with the power cord disconnected) can minimize ground-loop problems when making floating and common-mode measurements. Another way to minimize such problems is to connect the meter power cord to an isolation transformer (as discussed under "General Safety Precautions" in Sec. 3-1).

Resistance Measurements

The following considerations apply to the autoranging multimeter of Fig. 3-6. However, similar considerations apply to a variety of similar meters when measuring resistance.

Note that there are two resistance ranges, Ω_{LV} and Ω_{HV}. The Ω_{LV} range has a full-scale test voltage of 200 mV, whereas the Ω_{HV} range has a full-scale test voltage of 2 V. The threshold of conduction voltage of a silicon semiconductor (transistor, diode, etc.) is about 0.5 V. The threshold for a germanium semiconductor is about 0.2 V. Since the full-scale test voltage on Ω_{HV} exceeds the conduction thresholds on all types of semiconductors, erroneous (usually low) readings may result when resistance measurements are made in circuits containing diodes, transistors, or other semiconductors. Thus, the Ω_{LV} range should be used for resistance measurements in such circuits. The Ω_{HV} range is best suited for measurements that require a higher test voltage across the resistance, and for very high resistances up to 20 megohms, but should be avoided when semiconductors are included in the circuit. This subject is discussed further in Sec. 3-8.

The accuracy specification for resistance ranges is $\pm$ (0.10 percent of reading + 0.10 percent of range + 1 count). As an example, a display reading of 1.000 kilohms will have an uncertainty of $\pm$0.004 kilohm.

When measuring resistance, the voltage developed across the measured resistance is directly proportional to the current applied. For example, a reading of 1.000 kilohm corresponds to a voltage of 1.000 V on Ω_{HV}. The current applied by the meter is determined by the measurement range used. The test current for each range is as follows:

LV Range	HV Range	Test Current
.2 k	2 k	1.0 mA
2 k	20 k	0.1 mA
20 k	200 k	10 μA
200 k	2 M	1 μA
2 M	20 M	0.1 μA

As shown by the table, the test current on the higher range is quite low. When measuring in pure resistive circuits, the response time of the meter is less than 3 sec. If there is appreciable capacitance in the circuit under test, the time for the low currents to charge the capacitors may be quite long. For example, on the Ω_{HV} function, and for the 20-megohm range, it could take more than 1 min for an accurate reading to be displayed if the circuit capacitance is 1 μF. Higher capacitances produce even longer response times.

Over an extended period of operating time, there may be some variation in the shorted input zeroing on the Ω_{LV} function. The least significant (right-hand) digit may, because of ambient temperature changes, vary from a display of .0000 k Ω as a result. Periodically check the zeroing (HOLD mode on 2-M Ω range), and "touch up" if required. This problem can be almost entirely eliminated by using the Ω_{HV} function.

The ZERO control may be used as an offset adjustment to remove small, residual resistance readings (such as the test lead resistance when using the .2-k Ω range on Ω_{LV}). The ZERO control has an equal effect on all Ω_{LV} ranges and a decade-reduced effect on all Ω_{HV} ranges. Therefore, ZERO offset should be used for only the range being used (HOLD mode). Readjust the ZERO control when the offset measurements are complete. This is done by setting the RANGE switch to HOLD and connecting the test leads together. Adjust then ZERO control until the display indicates .000 M Ω.

When making measurements of very high impedance sources, as when required to measure resistance on the 20-M Ω range, the input circuit may be susceptible to noise. Voltage-producing noise fields on the test leads may change the display significantly. To keep errors to a minimum under these circumstances, keep the test leads as short as possible and twist them to minimize any "pickup" effects. It is good practice to twist the leads whenever possible, to equalize any stray radiation effects.

3-4 METER-PROTECTION CIRCUITS

Most modern meters are provided with some form of protection against overloading, accidentally connecting the wrong functions to a particular circuit (such as connecting the ohmmeter to a voltage), and similar occurrences. There are several types of meter-protection circuits. However, they are usually one of the following three types.

Fuse

A fuse can be inserted in the "common" line of the meter circuit to protect the movement, shunt, and multiplier resistors. A fuse is usually effective only against large surges of current. A delicate meter movement can be damaged even by small current surges, if they are beyond the maximum capability of the movement. Note that fuses have resistance values that affect the accuracy of the ohmmeter segment of a VOM. If the fuses are to be replaced, an *exact* replacement must be used.

Diodes

Diodes are used frequently in meter protection circuits. The most common uses are *input clamp diodes* and *meter-movement varistor diodes*.

A clamp diode resistor network can be placed across the input of a meter (VOM, electronic or digital) as shown in Fig. 3-8. When the meter is connected to a voltage greater than the bias voltages applied to the network, the diodes conduct and current flows through the resistors. This drops the input voltage to a safe level within the capability of the meter.

Varistor Diode

A varistor diode can be placed across the meter movement as shown in Fig. 3-9. A varistor diode is usually made of silicon and has a high forward resistance until a certain forward voltage is reached. At this voltage (usually a fraction of 1 V), the forward resistance drops to almost zero. Therefore, if the voltage across the meter movement increases to the dangerous level, the forward resistance of the diode drops to near zero, and all the current is passed through the diode. This type of diode can be added to most VOMs and is sold as a modification component or modification kit.

Relay

Some meters are provided with a relay that will open the movement circuit should there be an overload. A typical relay protection circuit is shown in Fig. 3-10. Note that the relay is actuated when contacts are closed by the pointer being driven off-scale in either direction. When the relay is actuated, the "common" line or the power line is opened. Once the relay has been actuated, it will remain tripped until it is reset by a mechanical reset button.

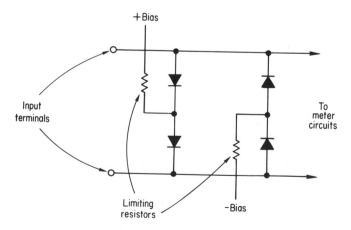

Fig. 3-8   Meter protection circuit using input clamp diodes.

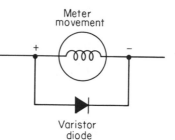

Fig. 3-9   Meter protection circuit using varistor
diode.

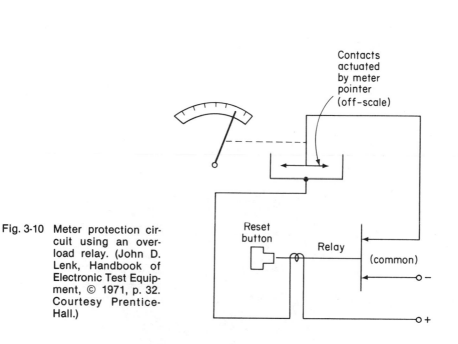

Fig. 3-10   Meter protection cir-
cuit using an over-
load relay. (John D.
Lenk, Handbook of
Electronic Test Equip-
ment, © 1971, p. 32.
Courtesy Prentice-
Hall.)

3-5 PARALLAX PROBLEMS

In nondigital meters, parallax is an error in observation which occurs when
the operator's eye is not directly over the pointer, as shown in Fig. 3-11.
This will cause the reading to appear at the right or left of the actual indication. Some manufacturers minimize this problem by placing a mirror
behind the pointer on the scale. Such a meter is shown in Fig. 3-12.

To use a mirrored scale most effectively, close one eye, and then move
the open eye until the pointer and its reflection appear to coincide, as shown
in Fig. 3-11.

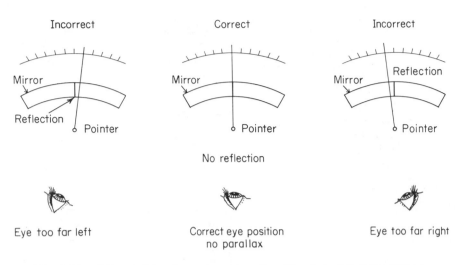

Incorrect   Correct   Incorrect

No reflection

Eye too far left   Correct eye position   Eye too far right

no parallax

Fig. 3-11   Using anti-parallax mirrored scales. (Courtesy B & K Precision.)

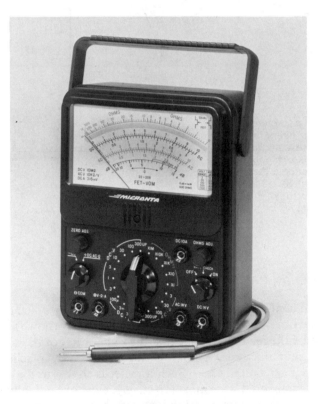

Fig. 3-12   Radio Shack FET-VOM with mirrored scales. (Courtesy Radio Shack.)

3-6 MOVEMENT ACCURACY PROBLEMS

No nondigital meter can be any more accurate than its basic movement. As the permanent magnet of a movement ages, the magnetic field weakens, and the indications will be in error (usually they will read low). This will be true even though the precision shunt and multiplier resistors may not have changed. Likewise, as a meter is subjected to shock, vibration, and overloads, the mechanical balance is disturbed, and the pointer moves from its zero position.

Meter movements are provided with maintenance adjustments of various types. Usually, both electrical and mechanical adjustments are provided.

Fig. 3-13 shows a typical meter movement electrical-compensation circuit. The purpose of this circuit is to provide a constant reading for a given voltage and current, despite any weakening of the movement magnet, loss of tension in the movement springs, and so on. Resistor R_1 is connected in shunt across the meter movement, and resistor R_2 is in series with the movement. (Note that both of these variable resistors are *internal* meter adjustments and are not operating controls. These resistor settings should not be touched except by an experienced meter technician.)

If the movement magnet weakens, shunt resistor R_1 is adjusted (increased) so that more current will flow through the movement coil for a given voltage. This will compensate for a low-reading movement. After adjustment of the shunt R_1, the series resistor R_2 is adjusted (decreased in this case) so that the *total resistance* of the meter movement circuit remains at the correct value, thus providing the correct reading for a given current.

Figure 3-14 shows two typical meter movements. Note that these movements are provided with a "mechanical-zero" adjuster. This adjust-

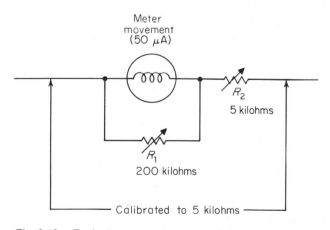

Fig. 3-13 Typical meter-movement electrical compensation circuit. (Courtesy Simpson Instruments.)

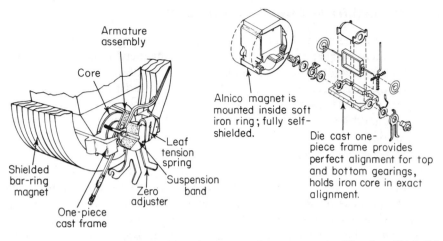

Labels on figure:

Armature assembly

Core

Alnico magnet is mounted inside soft iron ring; fully self-shielded.

Die cast one-piece frame provides perfect alignment for top and bottom gearings, holds iron core in exact alignment.

Leaf tension spring

Shielded bar-ring magnet

Suspension band

Zero adjuster

One-piece cast frame

Fig. 3-14   Suspension-type and bar-ring-type meter movements. (Courtesy Triplett.)

ment is used to set the movement to its *mechanical zero*, at which no current is flowing in the movement. The "mechanical-zero" adjustment is actuated by a screwdriver-type adjustment accessible from the meter front panel (shown directly below the meter face in Figs. 3-1, 3-2, and 3-12). This "mechanical-zero" adjustment is not to be confused with the "zero-ohm" adjustment or the "zero" adjustment of an electronic voltmeter. The mechanical adjustment permits the pointer to be set at zero *when no power is applied to the meter* and test leads are not connected.

3-7   READING ERROR PROBLEMS

There are many causes for inaccurate meter readings. Some of these are the result of "operator troubles," among them trying to obtain accuracy greater than the capability of the meter. However, it is possible to operate a meter properly and still obtain inaccurate readings. The following notes summarize the most common causes of reading errors.

The meter movement can be damaged, or the permanent magnet can be weakened, by excessive vibration or shock.

Overloads can damage meter movements, the multiplier and shunt resistors, and rectifiers (in the case of a-c meters). It is possible for an overload to damage the movement, resistors, or rectifier without actually burning out one or more of them. For example, excessive current through a resistor will cause heat. In turn, the heat can change the resistance value, making the shunt or multiplier out-of-tolerance.

On those meters which use printed-circuit wiring, an overload can cause breakdown between components or burn out part of the wiring board.

When ac is applied to a VOM set to measure dc, the ac will pass through the movement. This may or may not result in movement deflection, depending on meter construction, frequency of the a-c voltage, and so on. Even without deflection, it is possible for an a-c voltage to burn out or damage the meter. The presence of ac on a meter (when the meter is set to measure dc) will *sometimes* show up as vibration of the pointer.

Sometimes, there will be a buildup of static electricity on the meter face. This is often the result of wiping the face with a cloth. Such static electricity can cause the pointer to stick on the inside of the meter face and may show up as an incorrect reading.

If a meter is operated in the presence of a strong magnetic field, the moving coil may be deflected incorrectly. The permanent magnet may remain damaged even after the field is removed. (One particular problem in color-television service work is the presence of the degaussing coil. The strong a-c fields set up by the coil can completely destroy the permanent magnet in a meter movement.)

If a meter is operated in the presence of radio-frequency signals, the currents generated by the signals can cause reading errors. Electronic voltmeters and digital meters are most subject to such problems since both types of meters contain amplifiers that can amplify even small RF signals. Well-designed electronic voltmeters and digital meters are shielded against RF signals. The test leads are most vulnerable to RF signals. A shielded coaxial cable or a pair of shield test leads offer the best protection. If neither of these are practical, unshielded test leads can be twisted together to minimize the effects of RF signal pickup.

3-8 BASIC OHMMETER (RESISTANCE) MEASUREMENTS

The following paragraphs describe the steps necessary to make general ohmmeter (resistance) measurements with a VOM or electronic meter. Later chapters describe the procedures required to make specific resistance measurements, such as measuring leakage of a capacitor, front-to-back ratio of a diode, and so on.

The first step in making a resistance measurement is to zero the meter on the resistance range to be used. The meter can be zeroed on other ranges and on some meters will remain constant for all ranges. On other meters the ohmmeter zero will change for each range.

The meter is usually zeroed by touching the two test prods together and adjusting the "zero-ohms" or "ohms" control until the pointer is at "ohmmeter zero." This is usually at the right end of the scale for a VOM and at the left end for an electronic meter [see Fig. 3-15(a)].

Once the ohmmeter is zeroed, connect the test prods across the resistance to be measured [see Fig. 3-15(b)]. Read the resistance from the

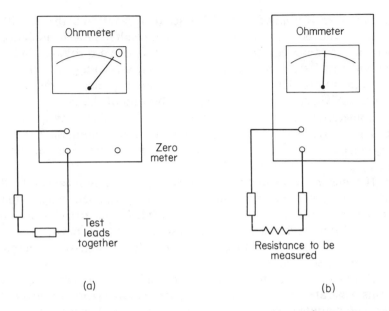

Fig. 3-15   Basic resistance measurement procedure.

ohmmeter scale. Make certain to apply any multiplication indicated by the range selector switch. For example, if an indication of "3" is obtained with the range selector at $R \times 10$, the resistance is 30 ohms. It should be possible to set the range selector to $R \times 1$ and obtain a direct reading of 30 ohms. However, it *may or may not* be necessary to zero the ohmmeter when changing ranges.

Two major problems must be considered in making any ohmmeter measurements. First, there must be no power applied to the circuit being measured. Any power in the circuit could damage the meter and cause an incorrect reading. Remember, capacitors often retain their charge after power is turned off. With power off, short across the circuit to be measured with a screwdriver to discharge any capacitance. Then make the resistance measurement.

Next, make certain that the circuit or component to be measured is not in parallel with (shunted by) another circuit or component that will pass direct current. For example, assume that the value of resistor R_1 in Fig. 3-16 is to be measured. If the battery in the ohmmeter were connected such that the diode CR_1 were forward-biased, current could flow through the transformer T_1 winding, diode CR_1, and choke coil L_1. All of these components have some d-c resistance, the total of which would be in parallel with resistor R_1.

If you suspect a reading when measuring resistance in a circuit con-

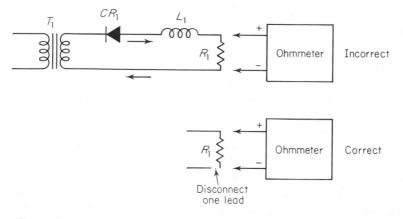

Fig. 3-16   Avoiding errors in resistance measurement due to parallel resistance.

taining any solid-state device (diode, transistor, etc.) that might pass current, try switching the leads. If the reading is substantially different with the
leads switched, look for a circuit component that has been forward-biased
by the meter voltage.

As discussed in Sec. 3-3, some meters are provided with high- and low-
voltage scales for resistance measurement. (The meter shown in Fig. 3-6 is
an example.) The low-voltage scales use a voltage that is below the conduction threshold of typical solid-state diodes or transistors. No matter what
type of meter is used, the simplest and safest method to eliminate parallel
resistance is to disconnect one lead of the resistance, as shown in Fig. 3-16.

3-9 BASIC VOLTMETER (VOLTAGE) MEASUREMENTS

The following paragraphs describe the steps necessary to make general voltmeter (voltage) measurements with a VOM or electronic meter. Later
chapters describe the procedures required to make specific voltage measurements, such as measuring the voltage at the terminals of a transistor.

The first step in making a voltage measurement is to set the range.
This is unnecessary for autoranging meters. Always use a range that is
higher than the anticipated voltage. If the approximate voltage is not
known, use the highest voltage range initially, and then select a lower range
so that a good midscale reading can be obtained.

Next, set the function selector to ac or dc as required. In the case of
dc, it may also be necessary to select either plus or minus by means of the
function switch. On simple meters, polarity is changed by switching the test
leads.

On an electronic voltmeter, the next step is to zero the meter. This should be done after the range and function have been selected. Touch the test leads together and adjust the "zero" control for a zero indication on the voltage scale to be used.

One common problem in any voltage measurement is that there may be both ac and dc in the circuit being measured. The following summarizes the various aspects of this problem.

If it is desired to measure ac only but dc is also present, the "output" or "a-c only" function can be selected, thereby switching a coupling capacitor into the input circuit. On a VOM, this is done by connecting the free test lead to the "output" terminal. In an electronic voltmeter, ac is often selected by means of a switch on the probe. In some meters, ac is always measured with a coupling capacitor at the input. In any event, the dc is blocked, and the ac is passed.

Use of the "output" function can present another problem. The coupling capacitor and the meter resistance form a high-pass filter and may attenuate low-frequency a-c voltages. However, most meters will provide accurate a-c indications above 15 or 20 Hz. It is also possible that the coupling capacitor and the meter movement coil will form a resonant circuit and increase the a-c signals at some particular frequency (usually about 30 to 60 kHz). Always consider the frequency problem when making any a-c voltage measurements.

If it is desired to measure dc only but ac is also present, there are several possible solutions. If the ac is of high frequency, it is possible that the meter movement will not respond and there will be no a-c indications when the meter is set to measure dc. If the a-c voltage is low in relation to the dc being measured, it is also possible that the meter will not be affected.

One solution, if the meter is affected by the presence of ac, is to connect a capacitor across the test leads. This will provide a bypass for the ac but will not affect the dc. However, the capacitor may affect operation of the circuit. Also, remember that the capacitor will be charged to the full value of the dc.

In some cases, it is possible to use a high-voltage or attenuator probe to measure dc in the presence of ac. The series resistance of the probe, combined with the natural capacitance between the probe's inner and outer conductor or shield, forms a low-pass filter. This filter action will have no effect on dc but will reject ac.

The fact that electronic voltmeters usually use some form of probe makes these instruments better suited to measure dc (in the presence of ac) than VOMs.

Remember that all voltage measurements (ac, dc, plus, minus, and decibels) are made with the meter in *parallel* across the circuit and voltage source, as shown in Fig. 3-17. This means that some of the current normally

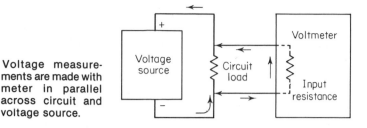

Fig. 3-17   Voltage measurements are made with
meter in parallel
across circuit and
voltage source.

passing through the circuit under test will be passed through the meter. In a VOM where the total meter resistance (or impedance) is low, considerable current may pass through the meter. This may or may not affect the circuit operation. For example, an oscillator that develops a small voltage over a high circuit impedance can be prevented from oscillating if a VOM is used to measure the voltage. (The low-impedance VOM draws excessive current, dropping the voltage to a point where oscillator feedback cannot occur.)

This problem of parallel-current drain does not occur in an electronic voltmeter, except when a voltage is measured across a high-impedance circuit. A typical electronic meter will have an input impedance of 10 to 15 megohms. If the circuit impedance is near this value, the current will divide itself between the circuit and the meter, possibly resulting in an erroneous reading.

3-10 BASIC AMMETER (CURRENT) MEASUREMENTS

The following paragraphs describe the steps necessary to make general ammeter (current) measurements. Later chapters describe the procedures required to made specific current measurements, such as measuring current drain on a power supply.

Most electronic voltmeters do not have a provision for measuring current, primarily because of their high input impedance. Since current must pass through the meter input circuit, there is a voltage drop across the meter. In an electronic meter, the voltage drop could be very high.

The first step in making a current measurement is to set the range. Always use a range that is *higher* than the anticipated current. If the approximate current is not known, use the highest current range initially, then select a lower range so that a good midscale reading can be obtained.

In many meters, selecting a current range involves more than positioning a switch. A typical VOM, such as shown in Fig. 3-12, requires that the test leads be connected to different terminals. In the case of the VOM of Fig. 3-12, the high-current range requires that the test leads be connected to the COM and d-c 10-A terminals. On all other current ranges, the COM and

V. Ω .A terminals are used. The range selector must be set to the appropriate range in all cases. Next, set the function selector to ac or dc as required. Most VOMs will not measure a-c current, so either plus or minus dc must be selected.

Note that when the lowest current scale is selected, such as 50 μA, the meter is actually functioning as a voltmeter. The meter movement is placed without a shunt in series with the circuit. Therefore, any sudden surges of current can damage the meter movement. This is especially a problem when there is both ac and dc in the circuit being measured. If the ac is of a higher frequency, it will probably have little effect on the meter movement. Lower-frequency ac can combine with the dc and possibly cause reading errors or meter movement burnout.

Remember that all current measurements (ac, dc plus, and minus) are made with the meter in *series* with the circuit and power source, as shown in Fig. 3-18. This means that all the current normally passing through the circuit under test will be passed through the meter. This may or may not affect circuit operation.

3-11 BASIC DECIBEL MEASUREMENTS

The procedures for measurement in decibels are similar to that for a-c voltage measurement, except: (1) the "output" function is always used for decibel measurements, and (2) the decibel scales are used instead of the a-c RMS or peak-to-peak scales.

When making decibel measurements, use the basic voltage measurement procedures of Sec. 3-9 and observe the precautions concerning decibel scales described in Sec. 3-2.

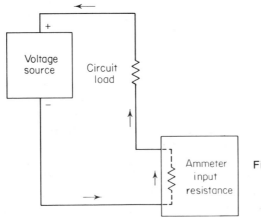

Fig. 3-18 Current measurements are made with meter in series with circuit and power source.

4

Testing and Calibrating Meters

All meters should be tested and calibrated periodically against known standards, since they can shift in accuracy due to damage or normal aging. Meters should also be checked out thoroughly when first placed in use by the technician. If the meters are new, they can be checked against manufacturer's specifications. If the meters are used and no specifications are available, the technician can establish a reference point for future periodic calibrations.

Laboratory meters must be checked against precision standards available in the laboratory or must be sent out to calibration (metrology) laboratories. A special problem is encountered with shop meters. Most shops are not equipped with precision standards, and it is expensive and time-consuming to send meters out for calibration. Therefore, means must be devised to check meters with available standards. The following procedures have been included to permit the calibration and testing of shop meters against commonly available standards and to make maximum use of these standards.

Note that these procedures are for various types of meters. If tests reveal that a meter is not up to standard, the instruction manual for the particular make and model of meter must be checked to find such information as the location of calibration controls or resistors and the actual calibration procedures.

4-1 OHMMETER TEST AND CALIBRATION

The accuracy of an ohmmeter can be checked by measuring the values of precision resistors. If the indicated resistance values are within tolerance, the ohmmeter can be considered as operating properly and ready for use. The following points should be considered when making ohmmeter accuracy checks.

The resistors should have a ± 1 percent or better rated accuracy tolerance. In any event, the resistor accuracy must be greater than the rated ohmmeter accuracy. A typical ohmmeter (VOM or electronic) will have a ± 2 or ± 3 percent accuracy.

Select resistor values that will provide midscale indications *on each ohmmeter range.* Make certain to zero the ohmmeter when changing ranges.

To determine *distribution error*, select precision test resistors that will give 25 and 75 percent scale indications in addition to the 50 percent (midscale) indication. Accuracy will not be the same on all parts of the scale due to nonuniform meter movements. However, accuracy should be within the rated tolerance on all parts of the scale and on all ranges.

Often ohmmeter accuracy will be rated in degrees of arc rather than a percentage of full scale. It is sometimes difficult to relate pointer travel in degrees of arc to a percentage. For practical work, remember that the ohmmeter accuracy is *approximately* equal to the accuracy of the d-c scale. For example, assume that the d-c scale is rated as accurate to within ± 2 small divisions (the scale has 100 small divisions with a ± 2 percent accuracy). Then the ohmmeter scale will also be accurate within the same degree of pointer travel, or arc, as it takes for the pointer to move ± 2 small divisions on the d-c scale.

4-2 VOLTMETER TEST AND CALIBRATION

Both the a-c and d-c scales of a voltmeter must be checked for voltage accuracy. In addition, the a-c scales must be checked for accuracy over the entire rated frequency range.

The obvious method to check the accuracy of a voltmeter is to measure a known voltage or series of voltages and check that the indicated voltage values are within tolerance. There are two basic methods for making such a test.

The most convenient method is to compare the voltmeter to be tested against a standard voltmeter of known accuracy. It is common practice in laboratory work to have one standard voltmeter, against which all meters are compared. The standard voltmeter is never used for routine work but only for test, and it is sent out for calibration against a *primary standard* at regular intervals.

The voltmeters to be tested are connected in parallel with the standard voltmeter and a variable-voltage source, as shown in Fig. 4-1. The source is then varied over the entire range of the voltmeters under test, and their voltage indications are compared with those of the standard voltmeter. In the case of a-c meters, the source is set to a given voltage; then the frequency is varied over the entire range of the meters under test.

In the type of test shown in Fig. 4-1, accuracy is dependent entirely on the accuracy of the standard meter and not on the source voltage or frequency. This is very convenient, since it is more practical to maintain a meter of known accuracy than a source of known accuracy, especially a source that can be varied over the entire voltage and frequency range of a typical VOM or electronic meter.

The next most popular method of checking a meter's voltage accuracy is to measure a voltage source of known accuracy. Usually, this voltage source is a *standard cell*. Inexperienced operators often check the voltmeter accuracy against a common dry cell or series of dry cells. This is satisfactory for rough shop work, but it is not accurate enough for precision work. A single dry cell rarely produces 1.5 V, as is often assumed. The accuracy of a dry cell voltage is usually less than that of a typical VOM.

A *mercury cell*, or series of mercury cells, provides much better accuracy than a common dry cell. The voltage output of a typical mercury cell is 1.35 V. The mercury cell will maintain this voltage for a long time, over a wide temperature range, and with an accuracy greater than that of a typical meter.

The most accurate source is a "standard cell" such as the *Weston cell*. There are two types of Weston cells: the *normal* cell and the *student* or *shop* cell. The normal cell is the most accurate, providing a voltage of 1.0183 V at 20 °C. The student or shop cell provides a voltage from 1.0185 to 1.0190 V at 20 °C. Therefore, the student cell can be off by as much as 0.5 mV. Usually,

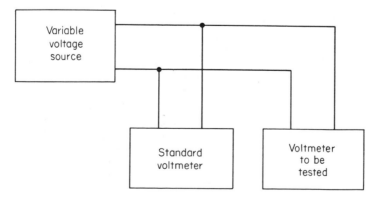

Fig. 4-1   Basic voltmeter calibration circuit.

this is greater accuracy than is required for all but precision laboratory meters.

Also, a student cell is less affected by changes in temperature. Once the accuracy of a student cell is established, it can be considered as remaining at that voltage over the normal range of room-temperature operation.

Inexperienced operators often connect a meter to be tested directly to a standard cell. Although this will provide an accurate indication, it will also place a damaging current drain on the standard cell. Laboratories generaly use some form of calibration circuit with their standard cells. Most of these calibration circuits use the *balance method* and require a galvanometer (refer to Sec. 1-2).

A calibration circuit using the balance method is shown in Fig. 4-2. Operation of the circuit is as follows. Switch S_1 is left in the open position. Potentiometers R_1 and R_2 are adjusted so that the voltage at the tap of R_1 is approximately equal to that of the standard cell. Switch S_1 is then set to the "standard cell" position, and the potentiometer R_1 is adjusted until the galvanometer reads zero. At this point, the voltage at the center terminals of S_1 is equal to the voltage of the standard cell. Initially, switch S_2 is open to place protective resistor R_3 in series with the galvanometer. Most galvanometers are quite sensitive. If there is a large voltage difference between the standard cell and the output of R_1, this difference could damage the galvanometer. Once approximate balance is reached, switch S_2 is closed, and the voltage is adjusted for exact zero on the galvanometer. Circuits of this type make it possible to obtain a balance while drawing less than 0.1 mA from the standard cell.

Once balance is obtained, switch S_1 is set to the "read" position. This

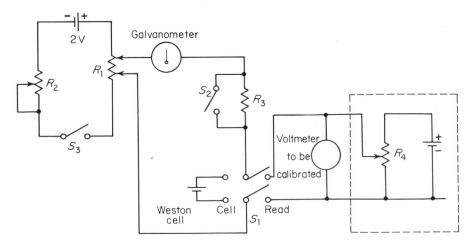

Fig. 4-2   Calibration circuit for d-c voltmeter.

removes the standard cell from the circuit and places the voltage across the meter under test. If a student or shop Weston cell were used in the circuit, the meter under test should read between 1.0185 and 1.0190 V.

A further variation of the calibration circuit is shown in Fig. 4-2 in dashed lines. This variation is used to protect the sensitive galvanometer. After the galvanometer circuit has been balanced against the standard cell and switch S_1 has been set to "read," resistor R_4 is adjusted until the galvanometer again shows a balance, indicating that the voltage at the tap of R_4 is equal to that of R_1 (and the standard cell). Then switch S_1 is set to "open," leaving only the voltage from R_4 across the meter under test.

It is obvious that the calibration circuit of Fig. 4-2 would provide a good indication on the 2.5 V scale of a meter, but it would be of little value on the 250-V range or any range substantially higher than 2.5 V. Likewise, the circuit could not be used at all on a lower range such as 250 mV. (Many of the newer meters have very low voltage ranges, which are necessary to measure the small voltage differences found in transistor circuits.)

These problems can be overcome by means of a *voltage-divider* circuit. The basic voltage divider can be any form of precision variable resistor from which the exact resistance can be read out or the exact ratio of full resistance to tap resistance can be shown by an external indicator. Such devices have various names, such as "volt box," "decade-ratio potentiometer," "ratiometer," or "decade box."

Voltage dividers can be used to provide precision voltage that can be calibrated against a standard cell, even though the voltages are much higher or lower than the cell voltage. For example, assume that a supposed 100-V source is placed across a precision variable-voltage divider of 100 ohms. Each 1-ohm tap or position on the voltage divider should then show a corresponding voltage (the 3-ohm position would show 3 V, the 33-ohm position would show 33 V, etc.). The divider could then be set to the 1.0185- or 1.0190-V position, and the resulting voltage balanced against a standard cell, as shown in the basic circuit of Fig. 4-3. This will establish the accuracy of the voltage source. For example, assume that the circuit shows a balance

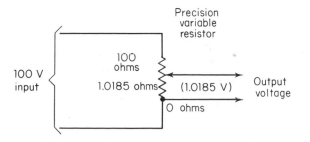

Fig. 4-3   Precision resistor used as a voltage divider.

against the 1.0190-V standard cell when the resistance shows 1.000 ohm, or approximately 2 percent off. This means that the voltage source across the voltage dividers is 2 percent off (2 percent high), and all the voltage readings will be 2 percent off. Note that the accuracy of the voltage-divider system depends on the accuracy of the voltage divider itself and not on the accuracy of the voltage source.

Now assume that a 2-V source is placed across a precision variable-voltage divider of 2000 ohms. Each 1-ohm position of the divider would then be equal to 1 mV with a 2000-m V total. The divider could be set to 1019 mV and balanced against a standard cell to determine accuracy. Once accuracy was set, the voltage divider could be used to check the entire 250-mV range of meters designed for transistor work.

The voltage-divider principle of Fig. 4-3 can be incorporated into the calibrating circuit of Fig. 4-2. Usually the voltage divider of Fig. 4-3 is substituted for the variable resistance R_1 of Fig. 4-2. With such an arrangement, the voltage across R_1 can be adjusted so that each position on R_1 is an accurate voltage source. For example, assume that R_1 is 2000 ohms and that an approximate 2-V source is placed across R_1 (adjusted by R_2). The voltage divider R_1 is set to indicate 1019 ohms, the standard cell is connected to the circuit by setting switch S_1 to "standard cell," and the galvanometer is balanced by adjustment of R_2 (which sets the total voltage across R_1). When the galvanometer is balanced, the voltage at R_1 is 1.0190 V. Switch S_1 is then set to "read," and the meter under test can be checked at any point from 0 to 2000 mV, as indicated by the resistance values of R_1.

4-3 AMMETER TEST AND CALIBRATION

The scales of an ammeter must be checked for accuracy of the current indication. A typical VOM will have only d-c current ranges, and a typical electronic meter will have no current ranges. Therefore, the primary concern is with d-c current accuracy. However, if a meter is provided with a-c current scales, they must also be checked for accuracy over the entire rated frequency range.

The most convenient method of checking current accuracy is to compare the ammeter to be tested against a standard ammeter of known accuracy. The ammeter to be tested is connected in series with the standard ammeter and a variable current source, as shown in Fig. 4-4. The source is then varied over the entire range of the ammeter, and the current indications are compared. The accuracy of this test is dependent entirely on the accuracy of the standard meter and not on the current source.

The next most popular method for checking current accuracy is to use a precision voltmeter and precision resistance. The ammeter to be checked

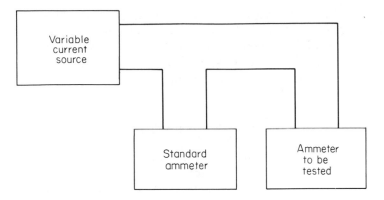

Fig. 4-4   Basic ammeter calibration circuit.

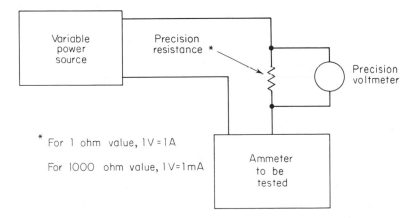

* For 1 ohm value, 1V = 1A

For 1000 ohm value, 1V=1mA

Fig. 4-5   Ammeter calibration circuit using a precision resistance and precision voltmeter.

and the precision resistance are connected in series with a variable power source, as shown in Fig. 4-5. The precision voltmeter is connected across the precision resistance. Current through the circuit is computed by Ohm's law ($I = E/R$).

The value of the precision resistor is chosen so that the voltage indicated on the precision voltmeter can be related directly to current. For example, if a 1-ohm resistance is used, the voltage across the resistance can be read directly in amperes (3 V equal 3 A, 7 V equal 7 A, etc.). If a 1000-ohm precision resistor is used, the voltage can be read directly as milliamperes (3 V equal 3 mA, etc.).

Note that the accuracy of this test method is dependent upon the accuracy of both the voltmeter and series resistance. The tolerance of both components must be added. For example, assume that both the voltmeter and resistance have a 1 percent tolerance. Then accuracy of the circuit could be no greater than 2 percent. In any event, the combined accuracy must be greater than that of the meter to be tested.

4-4   SHOP-TYPE TEST AND CALIBRATION

Most shops do not have a standard cell or even a standard meter of known accuracy against which the shop meters can be checked. The following procedures can be used to test and calibrate shop meters without the use of standards. These procedures are based on *comparison of internal scales*, such as comparison of the a-c scales against the d-c scales, or current scales against voltage scales, or one range against another. If the scales and ranges are compared with the *same source* and the scales or ranges agree within a given tolerance, it is usually safe to assume that the meter is accurate within that tolerance. Of course, these procedures will not reveal such conditions as a gradually weakening meter movement and should be used only for shop equipment or for a quick check of laboratory equipment. However, the procedures can be used to locate an inaccurate range or scale.

Comparison of D-C Voltage against D-C Current

1. Set the meter to measure d-c voltage. Connect the test leads to a fixed-voltage source and measure the voltage as shown in Fig. 4-6(a).
2. Set the meter to measure direct current. Connect the test leads to the *same* fixed-voltage source and a precision resistor, as shown in Fig. 4-6(b). Measure the current.
3. Using the *indicated*-voltage reading obtained in step 1, calculate the current by adding the precision resistance to the meter internal resistance (on that particular current range). Then divide by the indicated voltage.
4. If the indicated current agrees within tolerance with the calculated current, it is likely that both the voltage and current ranges are within tolerance. However, if the indicated current is quite different from the calculated current, either the voltmeter multiplier or the ammeter shunts is defective.

     For example, assume that a 3-V battery was measured in step 1 and that the voltmeter indicated 2 V due to a defective multiplier. Also assume that the total resistance (meter and resistor) is 3 ohms and that the current shunt is good. Using the 2-V figure for calculation (when

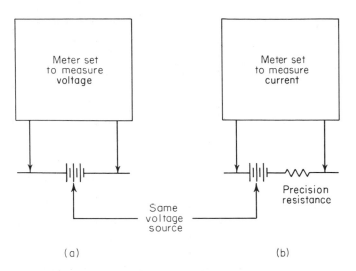

Fig. 4-6   Calibration of shop-type meters with voltage-current comparison.

the battery produces a true 3 V) will result in a lower calculated current. The calculated current would be 0.666 A, and the true indicated current would be 1 A.

5. It is important to make this test with the same power source for both voltage and current readings. Check each voltage range against each current range for a complete test.

6. It is also important to remember that the input resistance of the meter will change for each current range. The input resistances are usually listed in the meter instruction manual. If not, use the following procedure to determine the input resistance of a meter on each of its current ranges.

Finding Input Resistance of Current Ranges

1. Set the meter to measure direct current. Connect the test leads to a fixed-voltage source and precision resistor R_1, as shown in Fig. 4-7(a). The value of the resistor should be such that the current reading is approximately 25 percent of full scale.

2. Replace the precision resistor with another value (R_2) that will produce a reading 75 percent of full scale, as shown in Fig. 4-7(b). *Do not* change current ranges.

3. Using the following equation, calculate the input resistance of the current range:

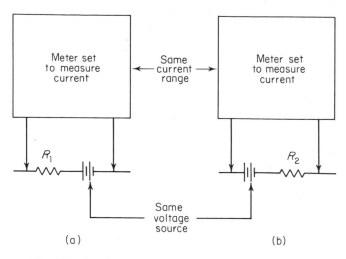

Fig. 4-7 Finding input resistance of current changes.

$$\text{input resistance} = \frac{I_1R_1 - I_2R_2}{I_2 - I_1}$$

where R_1 = resistance value that produces a 25 percent reading
       R_2 = resistance value that produces a 75 percent reading
       I_1 = current indicated when R_1 is in the circuit
       I_2 = current indicated when R_2 is in the circuit

4. Another method for determining input resistance of a current range is described in Chapter 5. However, this alternative method requires an additional voltmeter.

Comparison of A-C Voltage against D-C Voltage

1. Set the meter to measure d-c voltage. Connect the test leads to a fixed-voltage source and measure the voltage, as shown in Fig. 4-8(a). Read the d-c scales.

2. Without changing the test-lead connections to the voltage source and without changing the range, set the meter to measure a-c voltage. Measure the voltage as shown in Fig. 4-8(b). Read the a-c scale.

3. The a-c reading should be 1.11 times *higher* than the d-c reading (assuming that the meter uses a full-wave rectifier circuit). The reason for the difference in readings is as follows. The output of a full-wave rectifier is 0.637 times the peak voltage. The meter movement responds to this *average*. However, the ac scales of the meter read out in RMS values, or 0.707 times the peak voltage (0.707 = 1.11 × 0.637).

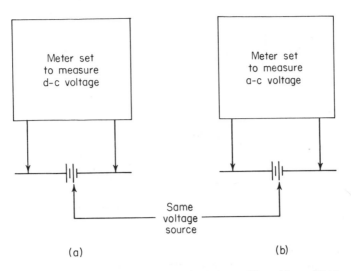

Fig. 4-8   Calibration of shop-type meters with a-c/d-c voltage
comparison.

4. The a-c reading will be 2.22 times *higher* than the d-c reading if the
   meter uses a half-wave rectifier, since the average output of a half-
   wave rectifier is 0.318 times the peak voltage (0.707 = 2.22 × 0.318).
   Note that most modern meters do not use a half-wave rectifier. A full-
   wave or full-wave bridge rectifier is used more often. The type of rectifier circuit can be determined by reference to the schematic or by
   visual inspection of the circuit. To make a quick test, try reversing the
   test leads connected to a d-c source with the meter set to measure ac. If
   the rectifier is half-wave, the pointer will reverse and show a negative
   indication. There should be no change in reading for a full-wave rectifier.

Checking Frequency Response of Shop Meters

There are a number of problems in checking frequency response of shop
meters, even though the procedure is quite simple. The meter is connected
to a signal generator capable of providing signals over the entire frequency
range of the meter and/or probe, as shown in Fig. 4-9(a) or 4-9(b). The
voltage output of the generator is set to provide a good midscale indication
on the meter. Then, without changing the voltage output, the generator frequency is varied over the full range of the meter.

A typical VOM has a range from about 15 Hz to 100 kHz, whereas an
electronic meter's range is from 15 Hz to 3 MHz. If an RF probe [as shown
in Fig. 4-9(b)] is used, the frequency range is extended to approximately 250
MHz. In any case, the voltage indication should remain constant (within a

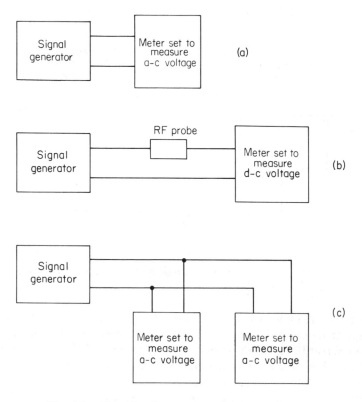

Fig. 4-9   Checking frequency response of meters.

given tolerance) over the entire frequency range. Each of the scales can be checked separately with different levels of voltage from the generator.

There are two basic problems with this procedure. First, shop generators rarely have a flat output over a wide frequency range. This may lead the inexperienced operator into false conclusions regarding the meter. The problem can be minimized but not completely eliminated by connecting the generator output to two meters simultaneously [in parallel, as shown in Fig. 4-9(c)].

If the generator output appears to change at some particular frequency on both meters, even if by different amounts on each meter, the generator output is not flat. For example, if the voltage output appears to drop at a particular frequency, the likelihood is greater that the generator output is dropping than both meters are giving an identical frequency response. (It is possible for two meters to have identical frequency responses, but it is not likely.)

Next, even if the output of a generator were flat over a given frequency range, it is likely that the meter components (movement inductance, multiplier resistance, and stray capacitance) will form filter circuits and resonant circuits at certain frequencies. These circuits can cause a false indication on the meter. A typical example occurs when the movement inductance and the stray capacitance combine to form a resonant circuit. This causes an apparent rise in voltage at the resonant frequency.

This problem can be minimized by measuring the same voltage on two ranges. Measure the voltage at the high end of one range, then change to the low end of the adjacent range (or vice versa) and check for the same voltage indication. The effects of stray capacitance and multiplier resistance are changed when meter ranges are changed. If the voltage indication changes, it is likely to be caused by a problem in the meter circuits. Note that there may be some vibration at low frequencies. This is typical for shop meters and is caused by inertia in the meter movement.

Also note that the input circuits of most electronic meters are provided with frequency-compensation capacitors. A typical circuit is shown in Fig. 4-10. The capacitor (usually in the order of 3 to 15 pF) is in parallel with the input resistance (typically 10 megohms). The capacitor is adjustable to provide a flat frequency response at the high end of the meter frequency range. As in the case of other internal meter adjustments, it is not recommended that the frequency-compensation capacitors be calibrated except by a qualified instrument technician.

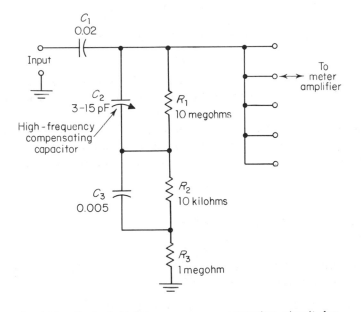

Fig. 4-10   Typical high-frequency compensation circuit for electronic meter.

Checking the frequency range of a *demodulator probe* requires a special test circuit as shown in Fig. 4-11. A demodulator probe produces both a d-c output (by rectifying the RF) and an a-c output (or pulsating d-c output) which is the modulation signal. The d-c output can be checked in the same way as any RF probe (Fig. 4-9). The a-c or demodulated output must be checked on the a-c scales of the meter. For a thorough test, the modulation frequency should be varied over the full range of the probe. This usually requires that the RF-signal generator be modulated by an external audio generator.

With the equipment connected as shown in Fig. 4-11, set the meter to measure a-c voltage. Set the signal generator to maximum voltage output at any frequency well within the RF range of the probe. If necessary, temporarily set the meter to measure d-c voltage and note the indication produced by the RF signal. Vary the frequency of the external audio generator over the demodulation range of the probe. Typically, a probe will demodulate signals up to about 20 kHz. However, the a-c voltage indications will usually start dropping off at about 17 kHz.

Checking Full-Wave Bridge-Rectifier Operation

If any problem is found when checking the a-c ranges of a shop meter and that problem (inaccuracy, erratic operation, etc.) appears on all ranges, the rectifier circuit is the most likely cause of the trouble. Most meters now use some form of full-wave rectifier. The rectifier diodes and circuit can be submitted to a quick check without opening the meter case.

1. Set the meter to measure a-c voltage.
2. Connect the test leads to a fixed d-c voltage source, as shown in Fig. 4-12(a). Use a range and voltage source that will give a good midscale indication.
3. Reverse the test leads as shown in Fig. 4-12(b). The voltage indication should be approximately the same as obtained in step 2. If there is a noticeable difference in voltage indication, one or more of the diodes may be defective.

Checking a Zero-Center Scale for Positive-Negative Deflection

The zero-center scale of an electronic meter can be checked in a manner similar to that of the full-wave rectifier.

1. Set the meter to measure d-c voltages on the zero-center scale.
2. Set the pointer to zero center with the "zero" control.

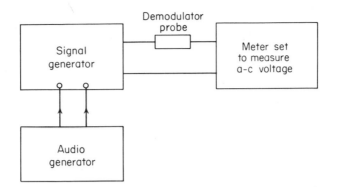

Fig. 4-11 Checking audio frequency range of demodu-
          lator probe.

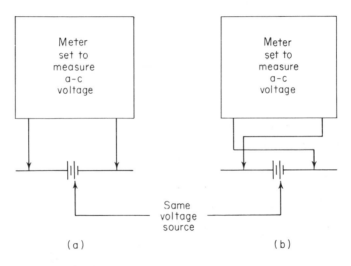

Fig. 4-12 Checking full-wave bridge rectifier operation.

3. Connect the test leads to a fixed d-c voltage source as shown in Fig. 4-13(a). Note thc indication on the zero-center scale. Usually, these scales are not calibrated in actual voltage but show divisions to the right or left of zero center.

4. Reverse the test leads as shown in Fig. 4-13(b). Note that the pointer moves in the opposite direction from zero center compared with what it did in step 3, but it does so by the same number of divisions as obtained then.

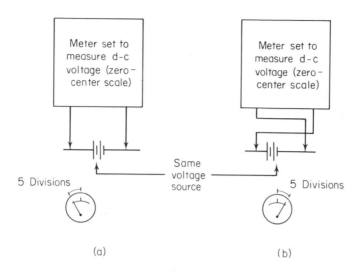

Fig. 4-13   Checking zero-center scale for positive-negative
deflection.

4-5 TESTING METER CHARACTERISTICS

In addition to testing a meter for accuracy on its resistance, voltage, and current scales, it is often helpful to check other characteristics of the meter circuit. The following paragraphs describe the procedures for testing the most important meter characteristics.

Checking Ohms-per-Volt Rating of a VOM

There are two methods for measuring the ohms-per-volt rating of a VOM. One method (with fixed resistor) will quickly determine if the ohms-per-volt characteristic is correct (as rated by the manufacturer) but will not show the actual ohms-per-volt rating. The second method (with variable resistor) will show the actual ohms-per-volt rating. The *fixed-resistor method* includes the following steps:

1. Set the meter to measure d-c voltage.
2. Connect the test leads to a fixed-voltage source [as shown in Fig. 4-14(a)] that will provide an approximate 75 percent full-scale reading.
3. Disconnect the test leads. Insert a precision resistor in the circuit as shown in Fig. 4-14(b) and reconnect the test leads. The value of the fixed resistor should be equal to the full-scale value multiplied by the

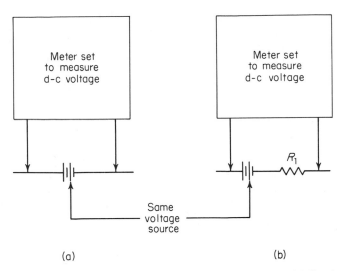

(a)   (b)

Fig. 4-14   Checking accuracy of ohms-per-volt rating with fixed resistor.

ohms-per-volt rating. For cxample, if the full-scale value is 3 V on a 20,000-ohms/V meter, the precision resistor value should be 60,000 ohms.

4. If the ohms-per-volt characteristics are correct, the voltage indication will drop to one-half that obtained in step 2.

5. Repeat the test on each of the meter scales, using precision resistors of the appropriate value for each scale.

6. The test can be madc on the a-c scales using the same procedure. However, the meter must be set to measure ac and an a-c source must be used. Also, it should be noted that the a-c ohms-per-volt rating is usually much lower than the d-c rating.

The *variable-resistor method* includes these steps:

1. Set the meter to measure d-c voltage.

2. Connect the test leads to a fixed-voltage source [as shown in Fig. 4-15(a)] that will provide an approximate 75 percent full-scale reading. Note the *exact* voltage reading.

3. Disconnect the test leads. Insert a variable resistor in the circuit as shown in Fig. 4-15(b) and reconnect the test leads. Set the variable resistor to the approximate value of the ohms-per-volt rating before connecting the resistor into the circuit.

4. Adjust the variable resistor until the voltage reading is exactly one-half that obtained in step 2.

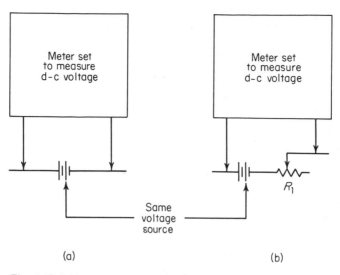

Fig. 4-15   Measuring ohms-per-volt rating with variable resistor.

5. Disconnect the test leads and measure the value of the variable resistor with a precision ohmmeter. The variable resistor's value is equal to the ohms-per-volt rating of the meter.

6. Repeat the test on all scales, including the a-c scales if desired.

NOTE

Note that the ohms-per-volt rating of any meter should be the reciprocal of the current required for a full-scale reading on the basic meter movement. For example, a 100-μA movement will produce a 10,000-ohms/V meter (1/0.0001 A = 10,000).

Checking Effect of "Output" Function on A-C Voltage Readings

As discussed in Sec. 3-9, use of the "output" function may cause an error in readings, particularly at low frequencies. The actual effect can be measured and a *scale factor* be applied to readings made in the "output" mode of operation.

1. Connect the meter to an audio generator as shown in Fig. 4-16.

2. Set the meter to measure a-c voltage directly and not through the "output" terminal.

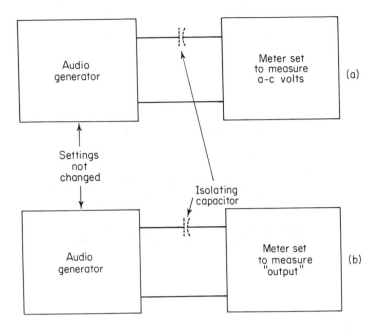

Fig. 4-16 Checking effect of "output" function on a-c voltage readings.

3. Set the audio generator to some low frequency near the low end of the meter-rated frequency range (20 Hz, 60 Hz, etc.).

4. Adjust the amplitude of the audio generator signal so that the meter indicates some a-c voltage near the midscale.

5. Without changing any of the audio generator or meter controls, apply the generator output through the "output" function.

6. If there is no change in voltage indication, the "output" function has no effect on the circuit. However, if the reading is lower in the "output" function (as is usually the case), a scale factor must be applied. For example, assume that the reading is 7 V on normal ac and 5 V in the "output" mode. Then a $\frac{7}{5}$ scale factor must be applied to all readings made on the "output" function. (Divide the "output" functioning reading by 5, then multiply by 7.)

7. If there is any d-c voltage at the audio generator output terminals, use an isolating capacitor as shown by the dashed line in Fig. 4-16.

5

Special Measurement Procedures

This chapter is devoted to special measurement procedures and techniques that can be applied to a number of components and circuits. Also included is information that will aid the experimenter in calculating the values of multipliers and shunts for meters.

5-1 CALCULATING MULTIPLIER AND SHUNT VALUES

It is possible to convert a basic meter movement into a multirange voltmeter (by adding multipliers) or ammeter (by adding shunts). To calculate the values for multipliers and shunts it is necessary to know the current required for full-scale deflection and the resistance of the basic meter movement. Usually, the full-scale deflection current is indicated on the scale face. However, the internal resistance is usually not marked.

Finding Internal Resistance of Meter Movements and Ranges

The internal resistance of a meter movement can be determined using the test circuits of Figs. 5-1 or 5-2. These same test circuits can be used to find the internal resistance of a VOM on its various d-c current ranges.

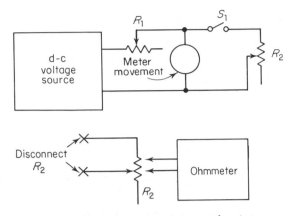

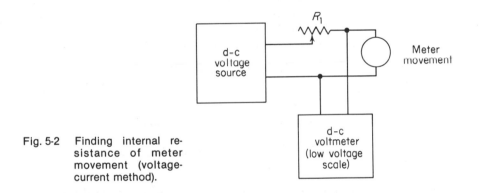

Fig. 5-1   Finding internal resistance of meter movement (half-scale method).

Fig. 5-2   Finding internal resistance of meter movement (voltage-current method).

To use the circuit of Fig. 5-1, disconnect R_2 from the circuit by opening switch S_1. Adjust R_1 until the meter movement is at full-scale deflection. Close switch S_1 and adjust R_2 until the meter movement is at exactly one-half scale deflection. Disconnect R_2 from the circuit and measure the R_2 resistance value with an ohmmeter. The R_2 resistance value is equal to the internal resistance of the meter movement.

To use the circuit of Fig. 5-2, adjust R_1 until the meter movement is at approximately 75 percent of full-scale deflection. Adjust R_1 for some exact current value. Calculate the internal resistance using Ohm's law ($R = E/I$). Note that the voltage drop across a typical movement will be less than 1 V. Therefore, it will probably be necessary to use the lowest scale of the voltmeter.

Never connect an ohmmeter directly across the meter movement. This will damage (and probably burn out) the movement.

131

Finding Multiplier Resistance Values

The multiplier resistance required to convert a basic meter movement into a voltmeter capable of measuring a given voltage range can be calculated using the following equation:

$$R_x = \frac{R_m (V_2 - V_1)}{V_1}$$

where R_x = multiplier resistance (in series with the meter movement)
R_m = internal resistance of the meter movement
V_1 = voltage required for full-scale deflection of the meter movement
V_2 = voltage desired for full-scale deflection (maximum voltage range desired)

For example, assume that a meter has a 50-μA full-scale movement (with 50 equal divisions on the scale), an internal resistance of 300 ohms, and that it is desired to convert the meter movement to measure 0 to 100 V. (Each of the 50 scale divisions will then represent 2 V.)

1. Find the voltage required for full-scale deflection of the meter movement: $E = IR$ or $(3 \times 10^2) \times (5 \times 10^{-5}) = 0.015$ V. This is voltage V_1.

2. Use the equation given above to find the value of R_x:

$$R_x = \frac{300(100 - 0.015)}{0.015} = 1,999,700 \text{ ohms}$$

3. To verify this multiplier resistance, add the meter internal resistance (300 ohms) and divide by the full-scale voltage obtained with the multiplier (100 V) to find the ohms-per-volt rating:

$$1,999,700 + 300 = \frac{2,000,000}{100} = 20,000 \text{ ohms/V}$$

4. Then divide the full-scale current (50 μA) into 1 to find an ohms-per-volt rating of 20,000. Both ohms-per-volt ratings should match.

Finding Shunt Resistance Values

The shunt resistance required to convert a basic meter movement into an ammeter capable of measuring a given current range can be calculated either of two ways.

With method 1, use the following equation:

$$R_x = \frac{R_m}{N - 1}$$

where R_x = shunt resistance (in parallel with the meter movement)
        R_m = internal resistance of the meter movement
        N = multiplication factor by which the scale factor is to be increased

For example, assume that a meter has a 50-μA full-scale movement (with 50 equal divisions on the scale), an internal resistance of 300 ohms, and that it is desired to convert the meter to measure 0 to 100 mA. (Each of the 50 scale divisions will then represent 2 mA.)

1. Find the multiplication factor (or N factor) by which the movement is to be increased: N equals the desired scale factor (0.1 A or 100 mA) divided by movement current (0.00005 A), or 2000.
2. Use the equation given to find R_x:

$$R_x = \frac{300}{2000 - 1} = 0.1508 \text{ ohm}$$

With method 2, follow these steps:

1. Find the voltage required for full-scale deflection of the meter movement: $E = IR$ or $(3 \times 10^2) \times (5 \times 10^{-5}) = 0.015$ V.
2. Subtract the meter movement full-scale current (0.00005 A) from the desired current (0.1 A) to find the current that must flow through the shunt: $0.1 - 0.00005 = 0.09995$ A.
3. Using Ohm's law, find the shunt resistance: $R = E/I$, or $0.015/0.09995 = 0.1508$ ohm.

5-2 FABRICATING TEMPORARY SHUNTS

The internal shunts of commercial meters are precision resistors. External shunts are usually strips or bars of metal connected directly between the meter movement terminals. Sometimes shunts are strips of metal mounted on insulators. Commercial shunts should be used for permanent operation. However, for *temporary use*, it is possible to extend the range of any basic meter movement may times using nothing more than a piece of wire.

The basic arrangement is shown in Fig. 5-3, and the calibration schematic is shown in Fig. 5-4. If a piece of wire is added across the terminals of a basic movement, part of the current will pass through the wire. If the wire is twisted as shown, it is possible to adjust the wire's resistance. This will control the amount of current through the wire. The wire is twisted or un-

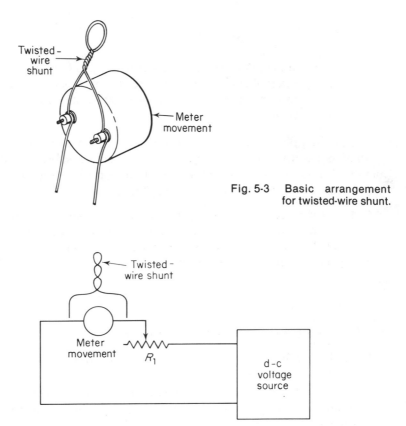

Fig. 5-3 Basic arrangement for twisted-wire shunt.

Fig. 5-4 Calibration circuit for twisted-wire shunt.

twisted as necessary, depending on whether more or less resistance is needed. It is not necessary to know the resistance of the shunt or the internal resistance of the meter movement but only the full-scale deflection of the movement and the full-scale deflection that is desired. To fabricate a twisted-wire shunt follow this procedure:

1. Connect the meter movement to the potentiometer and voltage source as shown. (Assume that the meter movement has a 50 μ-A full-scale rating.)

2. Set the potentiometer to its full value *before* connecting the meter. Then gradually reduce the potentiometer resistance until the meter reads full scale, or 50 μA.

3. Connect the twisted-wire shunt. The meter movement should drop back to zero.

4. Twist or untwist the wire until the meter movement reads exactly half scale, or 25 μA. (The actual current will still be 50 μA, since the potentiometer resistance is not changed.)

5. With the shunt wire connected and twisted so that an actual 50 μA flow will show as 25 μA (half scale), full scale will be 100 μA. Thus, the 50 μA movement has been extended to 100 μA.

6. If it is desired to extend the range further, adjust the potentiometer until the meter again reads full scale (which is now 100 μA). Then twist or untwist the shunt wire until the movement again reads half scale. The actual current flow will still be 100 μA, and the meter will be extended to a 200-μA full-scale rating.

7. In theory, the process could be repeated any number of times until the meter movement was extended to any desired range. In practice, a twisted-wire shunt has an obvious drawback. If the wire is exposed to any handling, the shunt resistance will change and throw the calibration off. But the method is quite accurate for *temporary use*.

5-3 EXTENDING OHMMETER RANGES

It is often convenient to extend the range of an ohmmeter to measure either very high resistances or very low resistances. This is possible using high-ohms or low-ohms adapters. The circuit for a typical high-ohms adapter is shown in Fig. 5-5. This basic circuit is suitable for use with a VOM or electronic meter. A low-ohms adapter suitable for a VOM is shown in Fig. 5-6. These adapters may be packaged as a form of probe, if desired. For greatest convenience, the adapter-circuit values should be chosen for a multiplica-

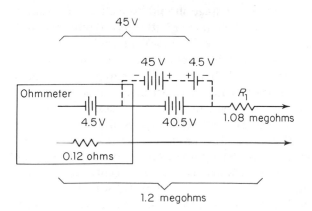

Fig. 5-5   High-ohms adapter circuit.

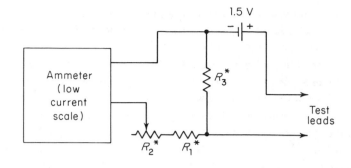

Fig. 5-6   Low-ohms adapter circuit.

tion factor of 10. That is, a high-ohms adapter should increase the ohm-meter reading 10 times, and a low-ohms adapter should decrease the reading 10 times.

Note that although both circuits use the ohmmeter scales for readout, the low-ohms adapter circuit does not use the VOM internal ohmmeter cir-cuit. Instead, the adapter circuit is connected to the lowest current scale (usually this means a connection directly to the meter movement).

High-Ohms Adapter

To increase the ohmmeter range by a factor of 10, operate the meter on its highest ohmmeter range (usually $R \times 10,000$ or $R \times 100,000$) and connect the circuit of Fig. 5-5.

The external-battery voltage should be such that the total voltage (ex-ternal battery plus internal ohmmeter battery) is 10 times that of the internal ohmmeter battery. For example, if a 4.5-V internal battery is used, the total voltage should be 45 V, and the external battery should be 40.5 V. If it is not practical to obtain a 40.5-V battery a 45-V battery should be used with a 4.5-V opposing battery in series. This arrangement is shown in dashed form in Fig. 5-5. The total voltage will then be the desired 45 V.

The value of the external resistor should be such that the total re-sistance (external resistor plus input resistance on the highest ohmmeter range) is 10 times that of the input resistance. A good approximation of the input resistance can be obtained by noting the center-scale indication on the ohmmeter. "12" and "4.5" are typical center-scale indications on VOMs. These represent 120,000- and 45,000-ohms input resistance, respectively, on the $R \times 10,000$ scale. Assuming that the input resistance is $120,000\,\Omega$, the total resistance would then be 1.2 megohms. This would require an external resistor of 1.08 megohms (1.1 megohms for practical purposes). The exact value of the external resistor is not too critical, since the entire circuit is adjusted for zero with the ohmmeter's "zero-ohms" control.

The high-ohms adapter is used in the same way as any conventional ohmmeter circuit. The test leads are shorted together, and the meter is zeroed with the "zero-ohms" control. Then measurements are made in the normal manner. To check the accuracy of a high-ohms adapter, measure the value of a precision resistor. The accuracy of the high-ohms circuit readings should be within 1 percent of readings made with the ohmmeter. Therefore, if the ohmmeter is rated at ±3 percent, then readings obtained with the high-ohms adapter connected should be ±4 percent.

Low-Ohms Adapter

To decrease the ohmmeter range by a factor of 10, operate the meter on its lowest current range (usually this is at the basic meter movement and will be 50 μA for a 20,000-ohms/V meter, 100 μA for a 10,000-ohms/V meter, etc.) and connect the circuit of Fig. 5-6.

The external-battery voltage is not critical. 1.5 V is chosen for convenience. However, the battery should have a high current capacity since there will be heavy current drain in the low-ohms circuit.

The value of a fixed resistor R_1 and external zero-adjust resistance R_2 combined in series should be approximately 10 times the internal resistance of the meter movement. The exact values of R_1 and R_2 are not critical since the entire circuit is adjusted for zero with R_2.

The value of shunt resistor R_3 should be *approximately* 0.1 times the center-scale ohms value when the ohmmeter is set to its lowest resistance range (usually $R \times 1$). For example, if the center-scale indication is 12 ohms (on $R \times 1$), the value of R_3 should be 1.2 ohms. A more exact value would be 0.095 times the center-scale reading, or 1.14 ohms for a 12-ohm indication.

The value of R_3 is critical, since the shunt resistor determines the 10 : 1 scale reduction. To check the accuracy of the low-ohms circuit, short the test leads together, zero the circuit with R_2, then measure the value of a precision resistor. Use a precision standard resistor in the range 3 to 7 ohms. Remember that a 3-ohm resistor will show a 30-ohm reading on the ohmmeter scale. Then try a 1-ohm (or less) precision resistor. If necessary, select a different value of R_3 to obtain a correct low-resistance reading. Make certain to zero the circuit with R_2 before making each resistance measurement. Also, do not keep the low-ohms adapter circuit connected to the circuit under test (or hold the test leads shorted) for a long time. This will cause considerable current to flow through R_3. The heat thus generated may change the resistance value of R_3. For best results, use a wire that is not heat-sensitive (such as Manganin wire).

The low-ohms adapter is used in the same way as any conventional ohmmeter circuit. The test leads are shorted together, and the meter is zeroed with the variable series resistance R_2. Then measurements are made in the normal manner. The low-ohms adapter is most effective in measuring values such as the resistance of a cold-soldered joint, the resistance of a switch contact, and so on.

5-4   EXTENDING VOLTMETER RANGES

The range of a voltmeter can be extended to measure high voltages by using a high-voltage probe, as discussed in Chapter 2, or by using *external* multiplier resistors connected to the basic meter movement, as shown in Fig. 5-7. In most meters, the basic movement is used on the *lowest* current range. Values for multipliers can be determined using the procedures of Sec. 5-1.

The basic meter movement can also be used to measure very low voltages. However, *great care* must be used not to exceed the voltage drop required for full-scale deflection of the basic movement. For example, assume that a 100-μA movement has a 2500-ohm internal resistance. Then the voltage drop required for full-scale deflection is 0.25 V. Voltages greater than this will destroy the meter. To be safe, make a rough check of the voltage to be measured on one of the meter's regular voltage scales. If the indication appears well below the full-scale deflection voltage of the basic meter movement, it is safe to proceed.

5-5   EXTENDING AMMETER RANGES

The range of an ammeter can be extended to measure high current values by using *external* shunts connected to the basic meter movement (usually the lowest current range), as shown in Fig. 5-8. Values for shunts can be determined using the procedure of Sec. 5-1.

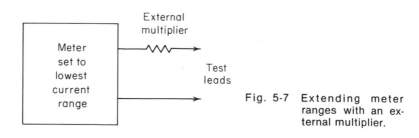

Fig. 5-7 Extending meter ranges with an external multiplier.

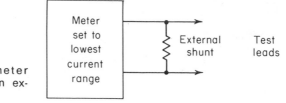

Fig. 5-8 Extending meter ranges with an external shunt.

Note that the range of the basic meter movement cannot be lowered. For example, if a 100-μA movement with 100 scale divisions is used to measure 1 μA, the meter will deflect by only one division. It is not practical to obtain greater deflection.

5-6 SUPPRESSED-ZERO VOLTAGE MEASUREMENTS

Sometimes it is convenient to measure a large voltage source that is subject to small variations on a low-voltage range. Thus, small voltage differences can be measured easily. For example, assume that a 100-V source is subject to 1-V variations (99 to 101 V). This would show up as one scale division on a 100-V scale. It is possible to use the suppressed-zero technique and measure the difference on a low scale, such as the 2.5- or 3-V scale found on most meters.

The basic suppressed-zero circuit is shown in Fig. 5-9. Note that an opposing voltage is connected in series with the voltage to be measured. Therefore, the meter sees only the *difference* in voltage. For example, assume that a 99- to 101-V source is to be measured on the 3-V scale of a VOM. A 98.5-V battery arrangement can be used as the opposing source. If the source to be measured is at 99 V, the meter reads 0.5 V (99 − 98.5). If the source is at 101 V, the meter reads 2.5 V (101 − 98.5). Both of these indications can be read easily on the 3-V scale.

Note that the suppressed-zero technique does not increase accuracy of voltage measurements but simply makes small voltage variations easy to read.

5-7 MEASURING COMPLEX WAVES

An oscilloscope is the best measurement for measuring complex waves since the waveshape, frequency, and peak-to-peak voltage can be found simultaneously. [The techniques for measuring complex waves with an oscilloscope are discussed in the author's *Handbook of Oscilloscopes: Theory and Application* (Englewood Cliffs, N.J.: Prentice-Hall, Inc., 1980)].

However, it is possible to measure the voltage of a complex wave with

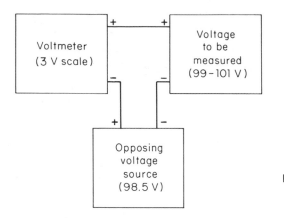

Fig. 5-9   Basic suppressed-
zero voltage meas-
urement technique.

a meter. The best results will be obtained from a peak-to-peak reading electronic meter, since peak-to-peak voltage is independent of actual waveform.
Other values of a complex wave (average, RMS, etc.) are dependent upon
the waveform (square wave, pulse, sawtooth, half-wave pulsating, full-wave
pulsating, etc.).

Most service manuals (such as those used in TV) specify the peak-to-
peak value of complex waves. For this reason, most electronic meters have
peak-to-peak reading scales.

Sometimes it is convenient to convert the peak-to-peak voltage of a
complex wave into RMS values. If the *waveshape is known*, the peak-to-
peak value can be converted into RMS using the data on Fig. 5-10. For example, assume that a sawtooth wave is measured with an electronic meter
and a peak-to-peak reading of 3 V is obtained. The RMS value of this
voltage is 0.8655 (3 × 0.2885).

It is also possible to obtain a reading of a complex wave with the RMS
scales of a meter. However, the indicated RMS value is not the true RMS
value (because the scales are based on the RMS of sine waves).

The RMS values obtained for complex waves on a conventional meter
can be converted to peak-to-peak readings using the data of Fig. 5-11. For
example, assume that a square wave is measured with a VOM and an RMS
reading of 3 V is obtained. The peak-to-peak value of this voltage is 5.4
(3 × 1.8). This value is developed as follows. The average value of a full-
wave rectified square wave is equal to the peak value (or one-half the peak-
to-peak value). A conventional RMS meter responds to average but indicates RMS, which is 1.11 times average in sine waves. Therefore, the
meter reads average (or peak in the case of a square wave) but indicates this
value as 1.11 times the peak. To find the peak, it is necessary to find the
reciprocal of 1.11, or 0.9. To find the peak-to-peak value, multiply the peak
value by 2 or 2 × 0.9 = 1.8.

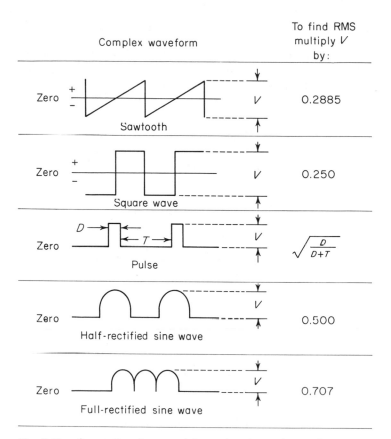

Fig. 5-10   Converting true peak-to-peak values of complex waves into RMS values.

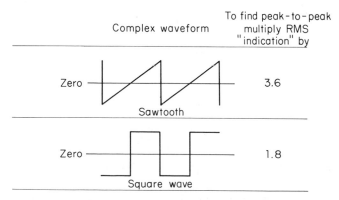

Fig. 5-11   Converting "indicated" values of complex waves on RMS meters into peak-to-peak values.

5-8 MATCHING IMPEDANCES

One problem often encountered when measuring complex waves is the matching of impedances. To provide a smooth transition between devices of different characteristic impedances, each device must encounter a total impedance equal to its own characteristic impedance. A certain amount of signal attenuation is usually required to achieve this transition. A simple resistive impedance-matching network that provides minimum attenuation is shown in Fig. 5-12 together with the applicable equations.

For example, to match a 50-ohm system to a 125-ohm system, we have $Z_1 = 50$ ohms and $Z_2 = 125$ ohms. Therefore,

$$R_1 = \sqrt{125\ (125 - 50)} = 96.8 \text{ ohms}$$

$$R_2 = 50\ \sqrt{\frac{125}{125 - 50}} = 64.6 \text{ ohms}$$

Though the network in Fig. 5-12 provides minimum attenuation for a purely resistive impedance-matching device, the attenuation as seen from one end does not equal that seen from the other end. A signal applied from the lower-impedance source (Z_1) encounters a voltage attenuation (A_1) that may be determined as follows. Assume that R_1 is 96.8 ohms and Z_2 is 125 ohms:

$$A_1 = \frac{96.8}{125} + 1 = 1.77 \text{ attenuation}$$

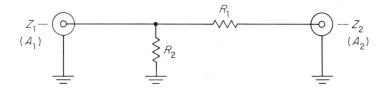

To match impedances: $R_1 = \sqrt{Z_2 (Z_2 - Z_1)}$

$$R_2 = Z_1 \sqrt{\frac{Z_2}{Z_2 - Z_1}}$$

Voltage attenuation seen from Z_1 end (A_1): $A_1 = \dfrac{R_1}{Z_2} + 1$

Voltage attenuation seen from Z_2 end (A_2): $A_2 = \dfrac{R_1}{R_2} + \dfrac{R_1}{Z_1} + 1$

Fig. 5-12   Resistive impedance-matching network.

A signal applied from the higher-impedance source (Z_2) will produce an even greater voltage attenuation (A_2) that may be determined as follows. Assume that $R_1 = 96.8$ ohms, $R_2 = 64.6$ ohms, and impedance $Z_1 = 50$ ohms:

$$A_2 = \frac{96.8}{64.6} + \frac{96.8}{50} + 1 = 4.44 \text{ attenuation}$$

Checking
Individual
Components

Meters are most often used to check components while the components are connected in their related circuits. However, it is possible to check many individual components using a VOM or electronic meter. These tests provide a quick check of the component's condition, usually on a "go/no-go" basis. In some cases, it is also possible to find the actual values or characteristics of components. This chapter describes the procedures for checking individual components out of circuit using conventional meters.

6-1 DIODE TESTS

For power rectifier diodes and small signal diodes, there are three basic tests required. First, any diode must have the ability to pass current in one direction (forward current) and prevent (or limit) current flow (reverse current) in the opposite direction. Second, for a given reverse voltage, the reverse current should not exceed a given value. Third, for a given forward current, the voltage drop across the diode should not exceed a given value.

In addition to the basic tests, a Zener diode must also be tested for the correct *Zener-voltage* point. Likewise, a tunnel diode must be tested for its *negative-resistance* characteristics. All these tests can be made with a meter. If the diode is to be used in pulse or digital work, the switching time must also be tested. This requires an oscilloscope and pulse generator. [A full discussion of diode testing is covered in the author's *Practical Semiconductor Data Book for Electronic Engineers and Technicians* (Englewood Cliffs, N.J.: Prentice-Hall, Inc., 1969)].

Diode Continuity Tests

A simple resistance measurement, or continuity check, can be used to test a diode's ability to pass current in one direction only. A simple ohmmeter can be used to measure the forward and reverse resistance of a diode. Figure 6-1 shows the basic circuit.

A good diode will show high resistance in the reverse direction and low resistance in the forward direction. If the resistance is high in both directions, the diode is probably open. A low resistance in both directions usually indicates a shorted diode.

It is possible for a defective diode to show a difference in forward and reverse resistance. The important factor in making a diode resistance test is the *ratio* of forward-to-reverse resistance (often known as *front-to-back* ratio). The actual ratio will depend upon the type of diode. However, as a rule of thumb, a small signal diode will have a ratio of several hundred to one, and a power rectifier can operate satisfactorily with a ratio of 10 : 1.

Diodes used in power circuits are usually not required to operate at high frequencies. Such diodes may be tested effectively with direct current or low-frequency alternating current. Diodes used in other circuits, even in audio equipment, must be capable of operation at higher frequencies and should be so tested.

Reverse-Leakage Tests

Reverse leakage is the current flow through a diode when a reverse voltage (anode negative, cathode positive) is applied. The basic circuit for measurement of reverse leakage is shown in Fig. 6-2. Note that a separate voltmeter and ammeter are required.

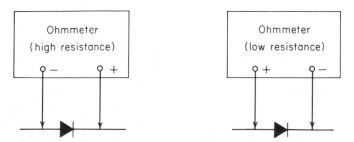

Fig. 6-1   Basic ohmmeter test of diodes for front-to-back ratio.

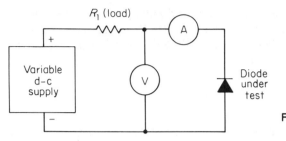

Fig. 6-2   D-c reverse leakage
test for diodes.

The diode under test is connected to a variable d-c source in the reverse-bias condition (anode negative). The variable source is adjusted until the desired voltage is applied to the diode as indicated by the voltmeter. Then the current (if any) through the diode is measured by the ammeter. This is the reverse (or leakage) current. Usually, any leakage current is undesired, but the limits should be determined by reference to the manufacturer's data sheet.

Forward-Voltage-Drop Tests

Forward voltage drop is the voltage that appears across the diode when a given forward current is being passed. The basic circuit for measurement of forward voltage is shown in Fig. 6-3. Note that a separate voltmeter and ammeter are required.

The diode under test is connected to a variable d-c source in the forward-bias condition (anode positive). The variable source is adjusted until the desired amount of current is passing through the diode as indicated by the ammeter. Then the voltage drop across the diode is measured by the voltmeter. This is the forward voltage drop. Usually, a large forward voltage drop is not desired. The maximum limits should be determined by reference to the manufacturer's data sheet. Typically, the forward voltage drop for a germanium diode will be approximately 0.2 V, and a silicon diode will have a forward drop of approximately 0.5 V.

Zener Diode Tests

The test of a Zener diode is similar to that of a power rectifier or small signal diode. The forward voltage drop test for a Zener is identical to that for a conventional diode. A reverse-leakage test is usually not required. In place of a reverse-leakage test, a Zener diode should be subjected to sufficient reverse voltage to determine the point at which avalanche occurs and reverse current begins to flow. If the external reverse voltage is increased beyond the avalanche point, additional current will flow and drop the

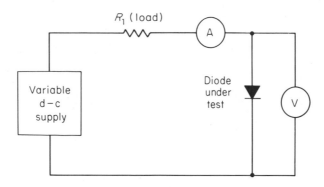

Fig. 6-3   D-c forward voltage-drop test for diodes.

voltage to a fixed level. This fixed level of voltage is called the Zener voltage.

The basic circuit for measurement of Zener voltage is shown in Fig. 6-4. As shown, the diode under test is connected to a variable direct-current source in the reverse-bias condition (anode negative). (This is the configuration in which the Zener is normally used.) The variable source is adjusted until the Zener voltage is reached and a large current is indicated through the ammeter. Zener voltage can then be measured on the voltmeter.

It is also common practice to test a Zener diode for impedance. However, the impedance check requires additional test equipment.

Tunnel Diode Tests

A tunnel diode must be tested for its *negative-resistance* characteristics. The most effective test of a tunnel diode is to display the entire forward-voltage and current characteristics on an oscilloscope. Thus, the valley and peak voltages as well as the valley and peak currents can be measured simultaneously.

It is possible to make a switching test of a tunnel diode using a meter. The basic circuit is shown in Fig. 6-5. As shown, the tunnel diode under test is connected to a variable d-c source. Initially the power source is set to zero; then it is gradually increased. As the voltage is increased, there will be some voltage indication across the tunnel diode. When the critical voltage is reached, the voltage indication will "jump" or suddenly increase. This indicates that the diode has "switched" and is operating normally. Usually, the voltage indication will be in the order of 0.25 to 1.0 V. The power source is then decreased gradually. With a normal tunnel diode, the voltage indication will gradually decrease until a critical voltage is reached. Then the voltage indication will again "jump" and suddenly decrease.

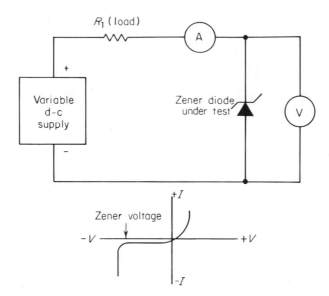

Fig. 6-4   Basic zener voltage test circuit.

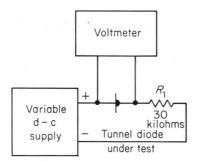

Fig. 6-5   Basic   tunnel-diode test circuit.

Diode Negative-Resistance Tests

Although Zeners and tunnel diodes are the only diodes normally operated in their negative-resistance region, some diodes may have negative-resistance characteristics. These characteristics can be tested using meters as shown in Fig. 6-6. Note that the diode under test is connected in the reverse-bias condition (anode negative). Therefore, any current indication on the ammeter is reverse current.

Initially, the power source is set to zero; then it is gradually increased until the voltage reading starts to drop (indicating that reverse current is flowing and the diode is in its negative-resistance region). The negative-

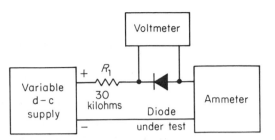

Fig. 6-6   Diode negative resistance test circuit.

resistance region should not be confused with leakage. True negative resistance is indicated when further increases in supply voltage cause an increase in current reading but a decrease in voltage across the diode.

The amount of negative resistance can be calculated using this equation:

$$\text{negative resistance (ohms)} = \frac{\text{decrease (volts) across diode}}{\text{increase (amperes) through diode}}$$

It is not recommended that a conventional diode be subjected to a negative-resistance test, unless there is a special need for the test. Also, do not operate a conventional diode in its negative-resistance region any longer than is necessary. Considerable heat is generated, and the diode may be damaged.

6-2   PHOTOCELL TESTS

There are two basic types of photocells: photovoltaic and photoconductive. The *photovoltaic cells* produce an output voltage and current in the presence of light and are often called *solar batteries* or solar cells. The *photoconductive cells* are often termed *light-sensitive resistors*, since they function as resistors and do not generate an output. Instead, photoconductive cells act as a resistance that varies in the presence of light, thus changing the amount of current being conducted through the circuit.

The basic test circuit for a photovoltaic circuit is shown in Fig. 6-7, and the photoconductive test circuit is shown in Fig. 6-8. In either circuit, the cell is exposed to sunlight or artificial light, and the meter reading is noted. Usually, a single cell will produce sufficient output to produce a readable indication on the low-voltage range of a VOM or electronic meter. However, some cells may require that the output must be read as current on a low-current range of the meter. For example, some cells produce less than 1 mA when exposed to strong sunlight. Always check the manufacturer's data for cell characteristics.

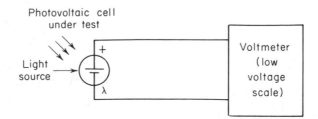

Fig. 6-7   Basic test circuit for photovoltaic cell.

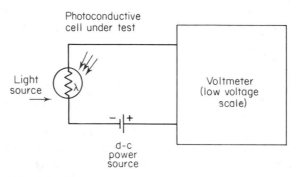

Fig. 6-8   Basic test circuit for photoconductive cell.

6-3 TRANSISTOR TESTS

There are four basic tests required for transistors in practical applications: gain, leakage, breakdown, and switching time. All of these tests are best made with commercial transistor testers and oscilloscopes. However, it is possible to test a transistor with an ohmmeter. These simple tests will determine whether the transistor is leaking and shows some gain. Usually, a transistor will operate in a circuit if the transistor shows some gain and is not showing any leakage (or leakage is very slight). [A full discussion of transistor testing is covered in the author's *Practical Semiconductor Data Book for Electronic Engineers and Technicians* (Englewood Cliffs, N.J.: Prentice-Hall, Inc., 1969)].

Transistor Leakage Tests

Transistors can be considered (for testing with a meter) as two diodes connected back to back. Therefore, each diode should show low forward resistance and high reverse resistance. These resistances can be measured with an

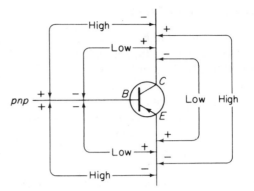

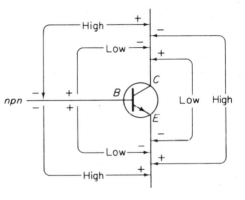

Fig. 6-9 Transistor leakage tests with an ohm-meter. (John D. Lenk, Handbook of Basic Electronic Trouble-shooting, © 1977, p. 161. Courtesy Prentice-Hall.)

ohmmeter as shown in Fig. 6-9. The same ohmmeter range should be used for each pair of measurements (base-to-emitter, base-to-collector, collector-to-emitter). On low-power transistors, there may be a few ohms indicated from collector to emitter. Avoid using the $R \times 1$ range or an ohmmeter with a high internal-battery voltage. Either of these conditions can damage a low-power transistor.

If both forward and reverse readings are very high, the transistor is open. Likewise, if any of the readings show a short or very low resistance, the transistor is shorted. Also, if the forward and reverse readings are the same (or nearly equal), the transistor is defective.

A typical forward resistance is 300 to 700 ohms. Typical reverse resistances are 10 to 60 kilohms. Actual resistance values will depend on ohmmeter range and battery voltage. Therefore, the ratio of forward to reverse resistance is the best indicator. Almost any transistor will show a ratio of at least 30 : 1. Many transistors show ratios of 100 : 1 or greater.

Transistor Gain Tests

Normally, there will be little or no current flow between emitter and collector until the base-emitter junction is forward-biased. This fact can be used to provide a basic gain test of a transistor using an ohmmeter. The test circuit is shown in Fig. 6-10.

In this test, the $R \times 1$ range should be used. Any internal battery voltage can be used, provided that it does not exceed the maximum collector-emitter breakdown voltage.

In position A of switch S_1, there is no voltage applied to the base, and the base-emitter junction is not forward-biased. Therefore, the ohmmeter should read a high resistance. When switch S_1 is set to B, the base-emitter circuit is forward-biased (by the voltage across R_1 and R_2), and current flows in the emitter-collector circuit. This is indicated by a lower resistance reading on the ohmmeter. A 10 : 1 resistance ratio is typical for an audio-frequency transistor.

Unijunction-Transistor Firing Test

The firing point of a unijunction transistor can be determined using a simple ammeter circuit. The test will also show the amount of emitter-base-one current flow after the unijunction is fired. If a unijunction transistor will fire with the correct voltage applied and draws the rated amount of current, it can be considered as satisfactory for operation in most circuits.

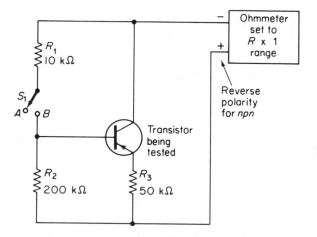

Fig. 6-10 Transistor gain test using an ohmmeter. (John D. Lenk, Handbook of Basic Electronic Troubleshooting, © 1977, p. 162. Courtesy Prentice-Hall.)

The test circuit is shown in Fig. 6-11. The base-two voltage is shown as + 20 V. However, any value of base-two voltage can be used to match a particular unijunction.

Initially, R_1 is set to zero volts (at the ground end). The setting of R_1 is then gradually increased until the unijunction fires. The firing voltage is indicated on the voltmeter. When the unijunction fires, the ammeter indication will increase suddenly. The amount of emitter-base-one current is read on the ammeter. Usually, the firing voltage is in the order of 0 to 20, and the current is less than 50 μA.

6-4. TRANSFORMER TESTS

A meter is the ideal instrument to check a transformer. The obvious test is to measure the transformer windings for opens, shorts, and the proper resistance values with an ohmmeter. If the ohmmeter is equipped with a high-ohms adapter (Chapter 5), it is possible to check a transformer for leakage between windings. In addition to basic resistance checks, one can test a transformer's proper polarity markings, regulation, impedance ratio, and center-tap balance with a voltmeter. If the transformer is tuned (such as an IF transformer), the Q factor can be checked with a meter. However, since a tuned transformer can be considered as a circuit, the procedure for measuring Q is described in Chapter 7 which covers circuit tests.

Checking Phase Relationships

When two supposedly identical transformers must be operated in parallel and the transformers are not marked as to phase or polarity, the phase relationship of the transformers can be checked using a voltmeter and a power source. The test circuit is shown in Fig. 6-12. For power transformers the source should be line voltage (115 V). Other transformers can be tested with lower voltage dropped from a line source or from an audio generator.

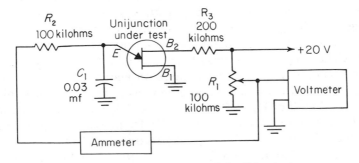

Fig. 6-11   Unijunction transistor firing test.

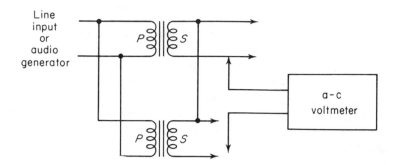

Fig. 6-12   Checking phase relationship of transformers.

The transformers are connected in proper phase relationship if the meter reading is zero or very low. The transformers are out of phase if the secondary-output voltage is double that of the normal secondary output. This condition can be corrected by reversing either the primary or secondary leads (but not both) of one transformer (but not both).

If the meter indicates some secondary voltage, it is possible that one transformer has a greater output than the other. This condition will result in considerable flowing of local current in the secondary winding and will produce a power loss (if not actual damage to the transformers).

Checking Polarity Markings

Many transformers are marked according to polarity or phase. These markings may consist of dots, color-coded wires, or some similar system. Unfortunately, transformer polarity markings are not always standard. Generally, transformer polarities are indicated on schematics as dots next to the terminals. When standards are used, the dots mean that if electrons are flowing *into* the terminal with the dot, the electrons will flow *out of* the secondary terminal with the dot. Therefore, the dots have the same polarity as far as the external circuits are concerned. No matter what system is used, the dots or other markings show *relative phase,* since instantaneous polarities are changing across the transformer windings.

From a practical standpoint, there are only two problems of concern: (1) the relationship of the primary to the secondary and (2) the relationship of markings on one transformer to those on another.

The phase relationship of primary to secondary can be determined using the test circuit of Fig. 6-13. First, check the voltage across terminals 1 and 3 and then across 1 and 4 (or 1 and 2). Then, assume that there are 3 V across the primary with 7 V across the secondary. If the windings are as shown in Fig. 6-13(a), the 3 V will be added to the 7 V and will appear as 10 V across terminals 1 and 3. If the windings are as shown in Fig. 6-13(b), the

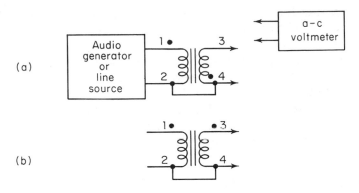

(a)

(b)

Fig. 6-13   Checking transformers for polarity or phase.

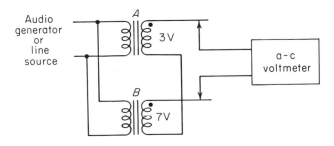

Fig. 6-14   Checking consistency of transformer polarity or phase markings.

voltages will oppose and will appear as 4 V (7 V − 3 V) across terminals 1 and 3.

The phase relationship of one transformer marking to another can be determined using the test circuit of Fig. 6-14. Assume that there is a 3-V output from the secondary of transformer *A* and a 7-V output from the secondary of transformer *B*. If the markings are consistent on both transformers, the two voltages will oppose and 4 V will be indicated. If the markings are not consistent, the two voltages will add, resulting in a 10-V reading.

Checking Transformer Regulation

All transformers have some regulating effect. That is, the output voltage of a transformer tends to remain constant with changes in load. Regulation is usually expressed as a percentage:

$$\text{percentage of regulation} = \frac{\text{no-load voltage} - \text{load voltage}}{\text{no-load voltage}}$$

Some transformers are designed to provide good regulation (a low percentage). Other transformers show very poor regulation (a high percentage).

Transformer regulation can be tested using the circuit of Fig. 6-15. The value of R_1 (load) should be selected to draw the maximum rated current from the secondary.

Only two steps are required for a test of regulation. First, measure the secondary-output voltage without a load and with a load. Then, use the equation given to determine percentage of regulation.

Checking Impedance Ratio

The impedance ratio of a transformer is the square of the winding ratio. (The impedance ratio should not be confused with the impedance of a transformer. Impedance measurements are discussed in Chapter 7.) If the winding ratio of a transformer is 15 : 1, the impedance ratio is 225 : 1. Any impedance value placed across one winding will be reflected onto the other winding by a value equal to the impedance ratio. For example, assume an impedance ratio of 225 : 1 and an 1800-ohm impedance placed on the primary. The secondary would then have a reflected impedance of 8 ohms. Likewise, if a 10-ohm impedance were placed on the secondary, the primary would have a reflected impedance of 2250 ohms.

Impedance ratio is related directly to the turns ratio (primary to secondary). However, turns-ratio data are not always available, so the ratio must be calculated using a test circuit as shown in Fig. 6-16. Use line voltage

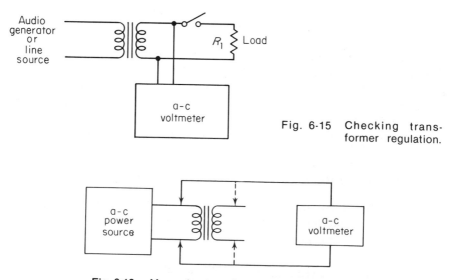

Fig. 6-15   Checking transformer regulation.

Fig. 6-16   Measuring transformer impedance ratio.

as a source for power transformers and an audio generator for other transformers. Measure both the primary and secondary voltages. Divide the
larger voltage by the smaller, noting which is primary and which is secondary. For convenience, set either the primary or secondary to some exact
voltage. The *turns ratio* is equal to one voltage divided by the other, and the
impedance ratio is the square of the turns ratio. For example, assume that
the primary shows 115 V with 23 V at the secondary. This indicates a 5 : 1
turns ratio and a 25 : 1 impedance ratio.

Checking Winding Balance

There is always some imbalance in center-tapped transformers. That is, the
turns ratio and impedance are not exactly the same on both sides of the
center tap. The imbalance is usually of no great concern in shop-type equipment, but it can be critical in laboratory-type transformers. It is possible to
find a large imbalance by measuring the d-c resistance on either side of the
center tap. However, a small imbalance might not show up, especially if the
d-c resistance is high.

It is usually more practical to measure the voltage on both sides of a
center tap, as shown in Fig. 6-17. If the voltages are equal, the transformer
winding is balanced. If a large imbalance is indicated by a large voltage difference, the winding should then be checked with an ohmmeter for shorted
turns or some similar failure.

6-5   RESISTOR TESTS

The obvious test of a resistor is to measure the resistance value with an ohmmeter. The various procedures for resistance measurement are discussed in
previous chapters. The following paragraphs describe the procedures for
measurement of special resistors and resistance elements.

Thermal or Ballast Resistors

Some resistors are made of materials that increase in resistance when the
temperature increases. Likewise, other resistors decrease in resistance value

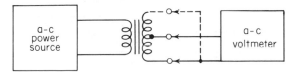

Fig. 6-17   Measuring   center-tapped   transformer
balance.

when the temperature increases. All resistors are subject to variation between "hot" and "cold" resistance values, but some resistors are manu-
factured specifically to produce a large change in resistance for changes in
temperature. The most practical way to test such a component is to heat the
resistor (either by operation in the circuit or by placing a voltage across the
resistor temporarily) and then measure the hot resistance. The cold resistance can be measured before heating the resistor, and the hot-versus-cold
resistance values compared.

Some resistors and resistance elements do not retain their "hot" resistance and must be measured while operating in a circuit. A vacuum-tube filament (or heater) is a good example. The usual tube heater will be less than 5
ohms cold and as high as 50 ohms hot.

The in-circuit value of a resistor can be found by measuring the current and voltage and then calculating resistance using Ohm's law ($R = E/I$).
The basic test circuit is shown in Fig. 6-18. Once the hot resistance is found,
the resistor can be removed from the circuit, and the cold resistance
measured with an ohmmeter.

A ballast resistor can also be checked using the same test circuit (Fig.
6-18). Most ballast resistors have a positive temperature coefficient
(resistance increases with temperature) to maintain constant current flow
even with changes in voltage. If voltage increases in a normal circuit, current flow will also increase. Under these conditions, a series-connected
ballast resistor will become hotter, its resistance will increase, and the current flow will be lowered. Also, voltage drop across the ballast resistor will
be treated, lowering the voltage to the load. Thus, the voltage and current
are maintained at a constant level.

In practice, there will always be some increase in current through a
ballast resistor with an increase in voltage. The amount can be checked using the test circuit of Fig. 6-18 and a variable power source. Increase the

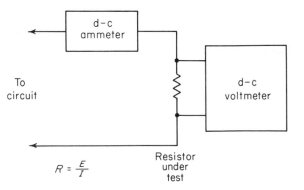

Fig. 6-18   Determining in-circuit "hot" resistance.

voltage until the current appears to level off. Then increase the voltage in small steps and note the change in current for each step. Be careful not to exceed the maximum rated voltage and current for the ballast resistor.

Checking Potentiometers and Variable Resistors

In addition to checking the resistance value, it is possible to check the contact of a potentiometer for "noise" or "scratchiness" using an ohmmeter. To do so, connect the ohmmeter between the wiper contact and one end of the winding. Then rotate the contact through the full range of resistance. The ohmmeter resistance reading should vary smoothly throughout the range. If the meter pointer "jumps" as the resistance is varied, the potentiometer contact is not riding firmly on the resistance winding or composition element. Often this condition is the result of dirt on the contact. A single application of contact cleaner will correct the problem.

6-6 CAPACITOR TESTS

The obvious test of a capacitor is to check for leakage with an ohmmeter. It is also possible to find the approximate value of a capacitor with an ohmmeter. Likewise, it is possible to check operation of a capacitor under various conditions with a voltmeter. The following paragraphs describe these procedures.

Checking Capacitor Leakage with an Ohmmeter

The capacitor must be disconnected from the circuit to make an accurate leakage test. The basic leakage test is similar to measuring any high-resistance value. The ohmmeter is connected across the capacitor terminals, and the resistance is measured. The following notes should be observed:

1. Use the highest resistance range of the ohmmeter. A typical capacitor will have a resistance in excess of 1000 megohms.
2. A more accurate test can be made if a high-ohms adapter (Chapter 5) is used. The higher voltage will show up any tendency of the capacitor to break down.
3. When using higher voltages, make certain not to exceed the voltage rating of the capacitor. This is especially a problem with the low-voltage electrolytic capacitors used in transistor radios and other solid-state equipment.
4. Another precaution to be observed when checking electrolytics is to make certain of the ohmmeter-battery polarity. One ohmmeter ter-

minal or lead is positive, and the other lead is negative. Usually, the positive terminal or lead is red, and the negative is black. However, to make sure, check the polarity against the meter schematic diagram or with an external meter connected to the ohmmeter leads.

5. Usually, capacitors will show some measurable resistance indication when the ohmmeter leads are first connected, but then the indication will increase to infinity. This temporary resistance indication is caused by current flow as the capacitor charges. If the resistance indication remains below 1000 megohms after the capacitor is charged, it is likely that the capacitor is leaking. On the other hand, if the resistance indication remains at infinity (the pointer never moves when the ohmmeter leads are connected to the capacitor), it is possible that the capacitor is open. Either of these conditions should be followed up with a further test.

6. Sometimes it is possible to charge a capacitor faster if the ohmmeter leads are first connected with the lowest range ($R \times 1$) in use. Then one should select each higher ohmmeter range, in turn, until the highest range is in use. (The lowest ohmmeter range applies the most voltage.)

Checking Capacitor Leakage with a Voltmeter

If a capacitor is suspected of leakage or of being open as the result of improper circuit operation or by the ohmmeter test just described, the facts can be confirmed using a voltmeter test.

If the capacitor is out-of-circuit, connect the capacitor leads across a voltage source and hold for approximately 10 sec. [see Fig. 6-19(a)]. For best results, use a voltage near the working voltage of the capacitor. Of course, never exceed the working voltage and always observe polarity for electrolytics.

With the capacitor charged, remove the voltage source and measure the *initial* capacitor voltage (with a voltmeter, *not an ohmmeter*). The initial voltage indication should be approximately the same as the source voltage [see Fig. 6-19(b)]. If no voltage is indicated, the capacitor is open. If the voltage is very low, the capacitor is leaking.

This test may not be too effective in testing low-value capacitors, especially with a VOM. The low input resistance of a VOM could discharge a low-value capacitor too quickly to produce a measurable voltage indication. However, an electronic meter will discharge the capacitor slowly because of the high input resistance.

If the capacitor is in-circuit, disconnect the capacitor ground lead and then measure d-c voltage from the lead to ground as shown in Fig. 6-20. Ini-

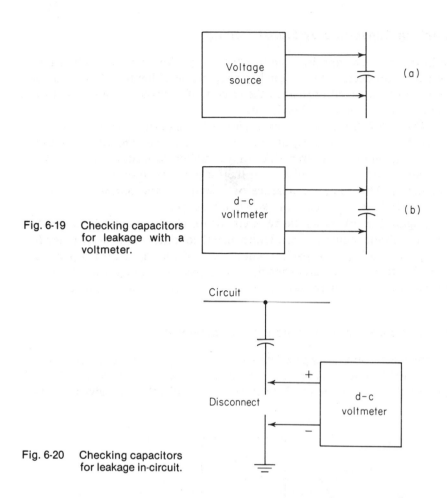

Fig. 6-19   Checking capacitors
for leakage with a
voltmeter.

Circuit

Disconnect

d-c
voltmeter

Fig. 6-20   Checking capacitors
for leakage in-circuit.

tially, there may be some d-c voltage indication due to capacitor charge. (The exact amount of initial indication will depend upon the input resistance of the meter and the capacitor value.) However, if the d-c voltage indication remains, the capacitor is leaking. It may be necessary to measure the initial charging indication on a high-voltage range of the meter. If the voltage indication drops, move the meter to the lowest voltage range to measure possible small voltages due to high-resistance leakage of the capacitor.

If the capacitor shows no voltage indication from the ground lead to ground, the capacitor is definitely not leaking. However, there is still the possibility of an open capacitor. A large-value capacitor will show an indication in most circuits. But, as in the case of the resistance test, a low-value capacitor may charge too quickly to produce a measurable reading.

Checking Capacitors by Signal Tracing

An in-circuit open capacitor can be checked quickly and positively using a voltmeter equipped with a signal-tracing probe (Chapter 2). Of course, there must be a signal present in the circuit. If necessary, connect a signal generator to the input of the circuit.

Figure 6-21(a) shows the basic circuit for checking a coupling capacitor. Under usual conditions, the a-c voltage (or signal) should be the same on both the input and output sides of a coupling capacitor. There may be some attenuation of the voltage on the output side (output will be lower than input). The complete absence of a signal at the output side of the capacitor indicates an open (or excessive leakage).

Figure 6-21(b) shows the basic circuit for checking a bypass capacitor (screen, cathode, emitter, etc.). Under usual conditions, the a-c voltage (or signal) *across* a bypass capacitor will be large if the capacitor is open. The usual function of a bypass capacitor is to pass a-c voltages or signals to ground. Therefore, there should be no a-c voltage across the capacitor.

Measuring Capacitor Values with a Voltmeter

It is possible to find the *approximate* value of a capacitor using a voltmeter. The method is based on the time constants of an RC circuit. A capacitor of a given value will charge to 63.2 percent of its full value in a given number

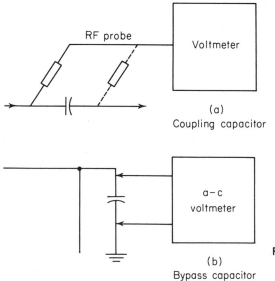

(a)
Coupling capacitor

(b)
Bypass capacitor

Fig. 6-21 Checking capacitors by signal-tracing.

of seconds through a given resistance value. Likewise, a capacitor will discharge to 36.8 percent of its full value through a given resistance in a given time. If the resistance value is known and the time measured, the capacitance value can be calculated. The basic discharge-method circuit is shown in Fig. 6-22, and the charge-method circuit is shown in Fig. 6-23.

Either a VOM or electronic voltmeter can be used. The input resistance of the meter must be known (Chapter 4 or 5). The high input resistance of an electronic voltmeter will cause the capacitor to charge and discharge slowly. Therefore, the electronic voltmeter should be used with small-value capacitors.

Use the following procedure with the *discharge circuit* of Fig. 6-22:

1. Move switch S_1 to the charge position and hold it for approximately 1 min.

2. Move switch S_1 to the discharge (meter) position. Simultaneously, start a stopwatch or note the exact time on the second hand of a conventional watch.

3. When the voltage has dropped to 36.8 percent of its initial value, stop

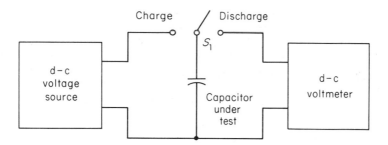

Fig. 6-22 Measuring capacitor values with a voltmeter (charge-discharge method).

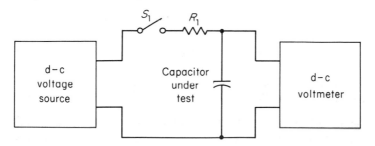

Fig. 6-23 Measuring capacitor values with a voltmeter (charge method).

the watch or note the elapsed time interval. For convenience, use an even voltage value such as 100 V, 10 V, and so on.

4. Divide the elapsed discharge time by the input resistance of the meter to find the capacitance in farads. If the input resistance of the meter is converted to megohms, the capacitance value will be in microfarads.

5. As an example, assume that the capacitor is charged to an initial value of 10 V, that the capacitor discharges to 3.68 V at the end of 90 sec, and that the input resistance of the meter is 3 megohms. $\frac{90}{3} = 30$ μF.

Use the following procedure with the *charge circuit* of Fig. 6-23:

1. Make certain that there is no voltage indicated across the capacitor from a previous charge.

2. Close switch S_1. Simultaneously start a stopwatch or note the exact time on the second hand of a conventional watch.

3. When the voltage reaches 63.2 percent of its final value, stop the watch or note the elapsed time interval. For convenience, use an even voltage value such as 100 V, 10 V, and so on. Also use an even R_1 resistance value, preferably in megohms so that the capacitance value can be expressed in microfarads. Note that there will be some voltage drop across R_1. Therefore, the capacitor will never fully charge to the value of the source voltage. Instead, the capacitor charge will depend upon the ratio of the resistance value of R_1 to the internal resistance of the meter. For example, if R_1 is 1 megohm, the meter is 3 megohms, and the source is 100 V, the capacitor will not charge beyond 75 V:

$$1 \text{ megohm} + 3 \text{ megohms} = 4 \text{ megohms}$$

$$\frac{100}{4} = 25$$

$$100 \text{ V} - 25 \text{ V} = 75 \text{ V}$$

4. Divide the elapsed charge time by the value of series resistance R_1 to find the capacitance in farads (if R_1 is in megohms the capacitance value will be in microfarads).

5. As an example, assume that the capacitor is charged to an initial value of 100 V (by a source of approximately 133 V), that the capacitor charges to 63.2 V at the end of 70 sec, and that the resistance of R_1 is 3 megohms. $\frac{70}{3} = 23.3$ μF.

Remember that these tests will provide an approximate capacitance value at best and will not be accurate with a capacitor that is leaking. However, the tests will show that the capacitor is operating normally and that its approximate value is correct.

Checking Capacitors during Troubleshooting

There are two basic methods for a quick check of capacitors during trouble-shooting. One method involves using the circuit voltages. The other technique requires an ohmmeter.

Checking Capacitors with Circuit Voltages As shown in Fig. 6-24(a), this method involves disconnecting one lead of the capacitor (the ground or "cold" lead) and connecting a voltmeter between the disconnected lead and

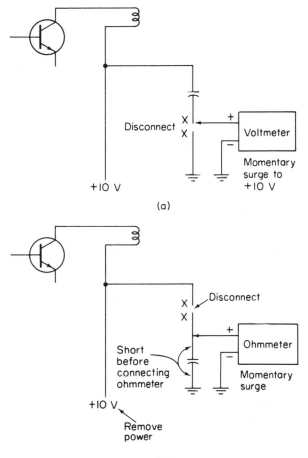

(a)

(b)

Fig. 6-24   Checking capacitors with circuit voltages (power applied) and with an ohmmeter (power removed). (John D. Lenk, Handbook of Basic Electronic Troubleshooting, © 1977, p. 166. Courtesy Prentice-Hall.)

ground. In a good capacitor, there should be a momentary voltage indication (or surge) as the capacitor charges up to the voltage at the "hot" end.

If the voltage indication remains high, the capacitor is probably shorted. If the voltage indication remains steady, but not necessarily high, the capacitor is probably leaking. If there is no voltage indication whatsoever, the capacitor is probably open.

Checking Capacitors with an Ohmmeter As shown in Fig. 6-24(b), this method involves disconnecting one lead of the capacitor (usually the "hot" end) and connecting an ohmmeter across the capacitor. Make certain all power is removed from the circuit. As a precaution, short across the capacitor to make sure that no charge is being retained after the power is removed. In a good capacitor, there should be a momentary resistance indication (or surge) as the capacitor charges up to the voltage of the ohmmeter battery.

If the resistance indication is near zero and remains so, the capacitor is probably shorted.

If the resistance indication is steady at some high value, the capacitor is probably leaking.

If there is no resistance indication whatsoever, the capacitor is probably open.

6-7 BATTERY TESTS

The obvious test for a battery is to measure the voltage from all the cells together or from each cell on an individual basis. Such a test will not show how a battery will maintain its voltage output under load. It is therefore necessary to test a battery under dynamic conditions. The following paragraphs describe these procedures.

Measuring Battery Output under Load

1. Connect the battery, load, and meter as shown in Fig. 6-25. If the battery is to be tested out-of-circuit, use a load resistance that will produce the maximum rated current flow.

2. Measure the battery voltage both without a load and with a load. Note any drop in voltage when the load is applied. If the battery is to be tested in circuit, measure the battery voltage with the equipment turned off. Then turn the equipment on and measure the full-load voltage.

3. Normally, there will be no more than a 10 or 15 percent drop in voltage output when a full load is applied. (The exact amount of voltage drop will depend upon the type of battery.) Also, the output

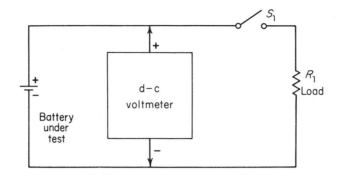

Fig. 6-25   Measuring battery output under load.

voltage of a good battery will return to full value when the load is removed. A defective battery (or cell) will drop in output voltage when a load is applied and will remain low.

4. A typical lead-acid storage battery will produce an output of approximately 2.1 V without a load and 1.75 V under full load. The condition of a storage battery can also be checked using a hydrometer (to measure specific gravity).

5. Always use the lowest practical voltage scale to measure battery voltage. This is necessary since an 0.1-V difference can be important in the single cell of a battery.

Locating a Defective Battery Cell

When it is necessary to locate a suspected cell in a group of many identical cells, connect all the cells in series across a load (Fig. 6-26). Remove the load. Then measure the voltage across each cell. The defective cell will show a lower output than the remaining cells or will possibly show zero output. In some cases, the polarity of the defective cell may reverse.

6-8 QUARTZ CRYSTAL TESTS

The *approximate* resonant frequency of a quartz crystal can be found using the test circuit of Fig. 6-27. To test a quartz crystal, use this procedure:

1. Set the meter to measure a-c voltage.

2. Adjust the signal-generator output to the supposed frequency of the crystal. Then adjust the signal-generator frequency for maximum indication on the meter. Read the crystal frequency from the signal-

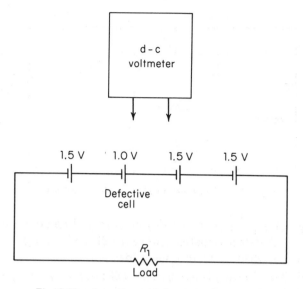

d – c
voltmeter

1.5 V   1.0 V   1.5 V   1.5 V

Defective
cell

R_1

Load

Fig. 6-26   Locating a defective battery cell.

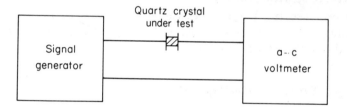

Quartz crystal
under test

Signal
generator

a – c
voltmeter

Fig. 6-27   Finding approximate resonant frequencies of
quartz crystal.

generator controls (frequency dial or counter). Be careful not to increase the signal-generator output voltage beyond the maximum rated limits of the crystal. Excess voltage can crack or otherwise damage the crystal.

7

Checking Circuit Functions

Meters are most often used to check circuit functions by measuring voltage, current, and resistance at various points in the circuit. This is the basis for all troubleshooting. If an expected reading is absent or abnormal, the related circuit components can then be checked on an individual basis. This chapter describes the procedures for checking *general* circuit conditions and the special problems that may arise. Chapter 8 describes the use of meters in checking specific circuits, such as amplifiers, communications equipment, television receivers, and so on.

7-1 MEASURING ALTERNATING CURRENT IN CIRCUITS

In-circuit current measurements are always a problem since the circuit must be interrupted (unless a clamp-on adapter is used). A-c measurements are a particular problem since most VOMs do not measure alternating current beyond 60 Hz, and electronic meters do not measure any form of current.

There are two basic solutions for the problem. Both ac and dc can be measured by inserting a low-value noninductive resistance in series with the circuit, measuring the voltage across the resistance, and then calculating the current using Ohm's law ($I = E/R$), as shown in Fig. 7-1. If a 1-ohm

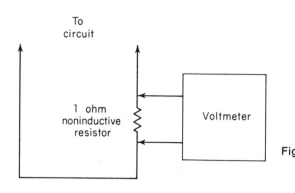

To
circuit

1 ohm
noninductive
resistor

Voltmeter

Fig. 7-1   Measuring in-circuit
current with resistor
and voltmeter.

resistor is used, the voltage indication can be converted directly into current (3 V = 3 A, 7 mV = 7 mA, etc.).

Alternating current can be measured by means of a transformer as shown in Fig. 7-2. The current to be measured is passed through the transformer primary. The voltage developed across the secondary and load is measured by the meter. There are commercial versions of this arrangement used in heavy power equipment. These are known as "current transformers" or "current adapters." In the commercial equipment, there are several taps on the primary winding. Each tap is designed to carry a particular current and produces a given voltage across the load. The commercial current-transformer circuit can be duplicated with individual components. However, because of the calibration problem (producing a specific voltage for a given current), it is more practical to use commercial equipment. Therefore, the method of measuring alternating current shown in Fig. 7-1 is recommended where commercial current adapters or transformers are not available.

7-2 MEASURING IMPEDANCE AND POWER CONSUMPTION IN A-C CIRCUITS

The current-measuring circuit of Fig. 7-1 can also be used to measure circuit impedance and power consumption. If the current and voltage are known, the impedance can be calculated by using the equation $Z = E/I$.

The power consumption (in volt-amperes) can be calculated by using the equation $VA = E \times I$. Of course true a-c power (expressed in watts) is found by multiplying the volt-amperes by the cosine of the phase angle. (The volt-ampered figure can be used for most practical purposes.)

Figure 7-3 shows the basic test circuit for measuring impedance. Note that the impedance of individual components (headphones, coils, transformers, etc.) or the impedance presented by a complete circuit or equipment (radio receiver, test equipment, etc.) can be measured using the same

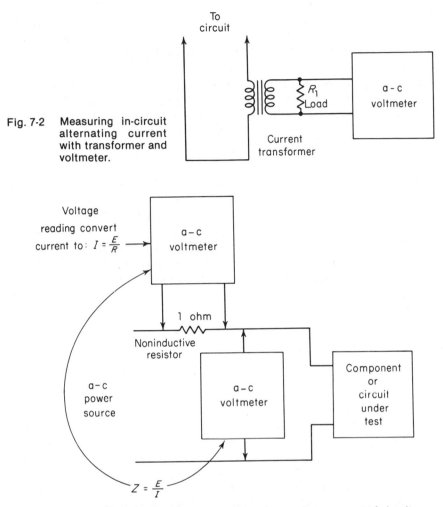

Fig. 7-2   Measuring in-circuit alternating current with transformer and voltmeter.

Fig. 7-3   Basic test setup for finding impedance of component of circuit.

basic circuit. The following precautions must be observed when using the voltage-current method to find impedance:

1. Accuracy of the measurements is dependent upon accuracy of the resistance value in series with the circuit.

2. A 1-ohm resistor should be used wherever possible. This will simplify the calculations and present the least disturbance to the circuit. If a 1-ohm resistor is not practical (because the voltage drop is too small to read), use a 1000-ohm or 10,000-ohm resistor. Remember that a 1000-ohm series resistor will convert voltage indications into milliamperes (7 V = 7 mA, etc.).

3. A noninductive series resistance must be used. If the resistor has any inductance (as do most wire-wound resistors) the resistor's inductive reactance will be added to the circuit and will produce an error in measurement.

4. The impedance found by the voltage-current test method applies only to the frequency used during the test. This is no problem for equipment operated from line power (radio receivers, test equipment, etc.), since the line-power frequency does not change. However, there is a problem for headphones, coils, transformers, and so on, that are operated over a wide range of frequencies. Therefore, these components should be tested over the anticipated frequency range. The component or circuit can be checked at various points throughout the frequency range, and any differences in impedance noted.

5. The composition of the waveform will also affect the impedance found using the voltage-current test method. For example, a nonsine wave or a sine wave containing many harmonics will produce a different impedance reading than will a pure sine wave. From a practical standpoint, it is usually not essential that circuits or components be tested for impedance with a pure sine wave. However, it is essential that the test be made with waveforms equivalent to those used in actual operation.

6. If the impedance and d-c resistance of a component or circuit are known, it is possible to find the reactance, inductance or capacitance, and power factor. These require considerable calculation or vector analysis and are usually of little practical value. However, the relationship between impedance and d-c resistance can be used to quickly determine the presence of reactance in a supposedly resistive load. For example, a T or L pad used in audio work is considered as a pure resistive load (at audio frequencies). It is possible that such a pad will present some reactive load, especially at the high end of the audio range (15 kHz and above). This condition can be determined by measuring the pad impedance at the audio frequency and then comparing the impedance against the d-c resistance. Both values should be substantially the same (allow for differences due to inaccurate measurements). If there is a large difference between the two values, there is some reactance present. (In the case of an L or T pad, it is probably inductive reactance.) An alternative impedance measuring procedure is given in Sec. 8-10.

7-3 MEASURING THREE-PHASE CIRCUITS

The procedures for measuring three-phase voltages and currents are essentially the same as for single-phase. However, it must be remembered that the voltage or current for each phase is not necessarily the same as for the line. Also, the output does not equal the input in a delta-to-star (or star-to-delta) transformer circuit. The important relationships in three-phase circuits are summarized in Fig. 7-4.

7-4 MEASURING POWER-SUPPLY CIRCUITS

The basic function of a power supply is to convert alternating current into direct current. This function can be checked quite simply by measuring the output voltage. In addition, it is often helpful to measure the regulating effect of a power supply, the power-supply internal resistance, and the amplitude of any a-c "ripple" at the power-supply output. The following paragraphs describe procedures for making such power-supply measurements using meters.

Measuring Power-Supply Regulation

Power-supply regulation is usually expressed as a percentage and is determined with the following equation:

$$\text{percentage of regulation} = \frac{\text{no-load voltage} - \text{full-load voltage}}{\text{full-load voltage}}$$

A low percentage-of-regulation figure is desired since it indicates that the output voltage changes very little with load. The following steps are used when measuring power-supply regulation:

1. Connect the equipment as shown in Fig. 7-5. Use a load resistance that will be equal to the operating load (or, where practical, use the operating load).
2. Measure the power-supply voltage without the load.
3. Apply the load and measure the power-supply voltage again.
4. Using the equation given, find the percentage of regulation. Note that power-supply regulation is usually poor (high percentage) when the internal resistance is high.

Measuring Power-Supply Internal Resistance

Inexperienced technicians often assume that the internal resistance of a power supply can be found by dividing the output voltage by the current. However, this figure is the *load resistance*, not the power-supply internal resistance.

Power-supply resistance is determined with the following equation:

$$\text{internal resistance} = \frac{\text{no-load voltage} - \text{full-load voltage}}{\text{current (amperes)}}$$

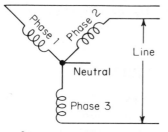

$$E_{phase}\ 1,2,\text{or}\ 3 = \frac{E_{line}}{1.73}$$

$$E_{line} = E_{phase}\ 1,2,\text{or}\ 3 \times 1.73$$

$$\text{Power} = E_{phase}\ 1,2,\text{or}\ 3 \times 1.73 \times$$

$$I_{phase} \times \text{Cosine}\ \theta$$

$$I_{line} = I_{phase}$$

Star, wye, or Y connections

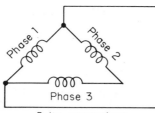

$$I_{phase}\ 1,2,\text{or}\ 3 = \frac{I_{line}}{1.73}$$

$$I_{line} = I_{phase}\ 1,2,\text{or}\ 3 \times 1.73$$

$$\text{Power} = I_{phase}\ 1,2,\text{or}\ 3 \times 1.73 \times$$

$$E_{phase} \times \text{Cosine}\ \theta$$

$$E_{line} = E_{phase}$$

Delta connections

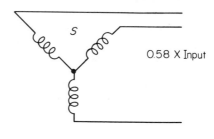

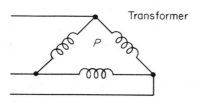

0.58 X Input

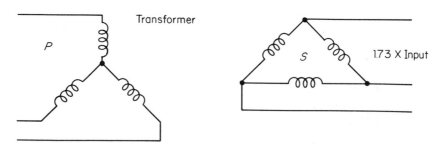

1.73 X Input

Fig. 7-4   Summary of three-phase voltage and current measurements.

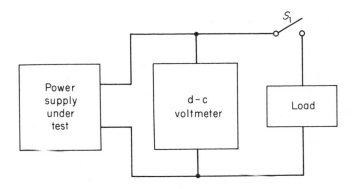

Fig. 7-5   Measuring power supply regulation.

For example, if the no-load voltage is 100, the full-load voltage is 90, and the current is 500 mA, then $100 - 90/0.5 = 20$ ohms.

A low internal resistance is the most desired since it indicates that the output voltage will change very little with load. To measure power-supply internal resistance, use this procedure:

1. Connect the equipment as shown in Fig. 7-6. Use a load resistance that will be equal to the operating load (or, where practical, use the operating load).

2. Measure the power-supply voltage without the load.

3. Apply the load and measure the power-supply voltage again.

4. Measure the load current.

5. Using the equation given, find the internal resistance.

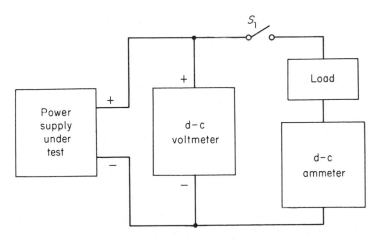

Fig. 7-6   Measuring power supply internal resistance.

175

Measuring Power-Supply Ripple

Any power supply, no matter how well regulated and/or filtered, will have some ripple. Measuring ripple is quite simple. Set the meter to measure ac only. For a VOM this means selecting the "output" function. For an electronic voltmeter, a series capacitor is inserted automatically when ac is selected. Any voltage measured under these conditions will be a-c ripple. Usually, the factor of most concern is the ratio between ripple and full d-c output voltage. For example, if 3 V of ripple is measured, together with 100-V d-c output, the ratio would be $\frac{3}{100}$ (or 3 percent).

One problem often overlooked in measuring ripple is that any ripple voltage (single-phase, three-phase, full-wave, half-wave, etc.) is not a pure sine wave. Most meters provide accurate a-c voltage indications only for pure sine waves. The service manuals for power-supply equipment often take this fact into account. However, some manufacturers specify peak-to-peak ripple voltage values.

7-5 MEASURING PAD CIRCUITS

L and T pads, whether considered as circuits or components, require special measurement procedures. Generally, pads can be checked using an ohm-meter. However, if it is suspected that a pad is producing some reactance (inductive or capacitive), the pad must be checked with an audio-signal source.

Measuring L Pads for Impedance

Usually, the input impedance of an L pad will remain constant, but the output impedance will vary as the pad setting is changed. (Or the pad can be reversed so that the output impedance is constant with the input impedance varying.)

The input-impedance test circuit is shown in Fig. 7-7. The load resistance R_L should be equal to the normal load, as seen from the output side of the pad. Vary the pad throughout its entire range and note the resistance values obtained at the input circuit. Assuming no reactance, the d-c input-resistance values will equal the input-impedance values.

The output-impedance test circuit is shown in Fig. 7-8. The load resistance R_L should be equal to the normal load as seen from the input side of the pad. Vary the pad throughout its entire range and note the resistance values obtained at the output circuit. Assuming no reactance, the d-c output-resistance values will equal the output-impedance values.

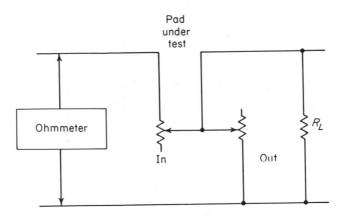

Fig. 7-7   Measuring L-pad input impedance.

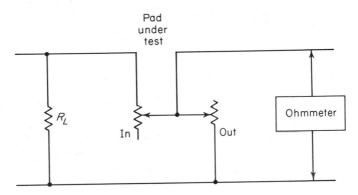

Fig. 7-8   Measuring L-pad output impedance.

Measuring T Pads for Impedance

The input and output impedance of a T pad should remain constant as the pad setting is changed (within a specified tolerance). This is one of the advantages of a T pad over an L pad. Note, however, that the input and output impedances of a T pad are not necessarily the same. Therefore, a T pad can also be used to match impedances as well as to vary signal strength.

The impedance test circuit is shown in Fig. 7-9. In Fig. 7-9(a), to measure input impedance, the load resistance R_L (equal to the normal load as seen from the output side of the pad) is connected at the output with the ohmmeter at the input. The circuit is reversed in Fig. 7-9(b) to measure output impedance.

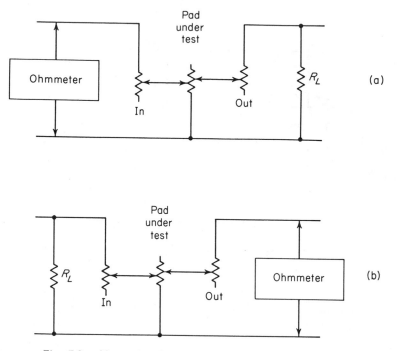

Fig. 7-9   Measuring T-pad input and output impedances.

In either circuit, vary the pad throughout its entire range and note that the resistance remains constant (within tolerance). Assuming no reactance, the resistance values will equal the impedance values.

Measuring Pads for Reactive Effect

An L or T pad can be checked for reactive effect using the circuit of Fig. 7-10. The load resistance R_L should be equal to the rated output impedance of the pad (or the impedance that is seen from the output side). The input resistance R_{in} should be equal to the rated input impedance of the pad. However, if the pad impedance is close to that of the audio-generator output impedance, then R_{in} should be added to the generator impedance to equal the pad input impedance. For example, if the pad input impedance is 10 Kilohms and the audio-generator output impedance is 50 ohms, R_{in} should be approximately 10 Kilohms. If the pad input is 100 ohms with a 50-ohm generator output, R_{in} should be 50 ohms. *The following steps should be adhered to when measuring L or T pads for impedance:*

1. Adjust the audio-generator output voltage for a good reading on the meter.

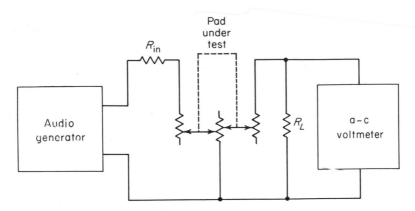

Fig. 7-10   Measuring pads for reactive effect.

2. Vary the audio-generator output over the entire frequency range with which the pad will be used. *Do not* change the generator output *voltage.*

3. If the pad has any reactive effect, the meter reading will not be constant. The meter reading will rise (or fall) at some particular frequency or over some particular frequency range.

4. Repeat the test at various settings of the pad. One point to remember when making this test or any test that involves varying the output of a generator: all generators do not have a flat output (constant-voltage output over the entire frequency range). The same is true of meters (Chapter 4). This can lead to an error in judgment. The ideal remedy is to calibrate the generator over its entire frequency range and record any variations. When this is not practical, a quick check can be made by connecting the meter directly to the generator output, varying the generator over the entire frequency range, and then noting any variations in output voltage. These variations can then be compared with any variations found with the meter connected at the pad output. If the variations are the same, it is likely that the generator or meter is at fault, not the pad.

7-6 MEASURING INTERNAL RESISTANCE OF CIRCUITS

It is sometimes convenient to measure the internal resistance of a circuit while the circuit is operating. For example, it may be desired to measure the plate resistance of a tube or the collector resistance of a transistor. Obviously, the circuit resistance cannot be measured with an ohmmeter while the

179

circuit is energized. However, it is possible to measure circuit resistance using a voltmeter and potentiometer. To do this, use the following procedure:

1. Connect the equipment as shown in Fig. 7-11.
2. Set the potentiometer to zero and measure the circuit voltage. This will be the full circuit voltage.
3. Increase the potentiometer resistance until the voltage is one-half the full circuit voltage (one-half that obtained in step 2).
4. Disconnect the potentiometer from the circuit. Measure the d-c resistance of the potentiometer.
5. Subtract the input resistance of the meter from the d-c resistance of the potentionmeter. The remainder is equal to the internal resistance of the circuit.

For example, assume that the meter reads 100 V when the potentiometer is set to zero and 50 V with the potentiometer set to 100,000 ohms. Also assume that the input resistance of the meter is 30,000 ohms. The circuit internal resistance is 70,000 ohms.

7-7 MEASURING VOLTAGE-SENSITIVE CIRCUITS

Many circuits are voltage-sensitive. That is, the voltage will vary with changes in load. These circuits are difficult to measure since the load presented by the meter will change the voltage from the normal operating value. Since a VOM has a lower input resistance than an electronic voltmeter, the VOM presents a greater load on the circuit and thus changes the voltage by a greater amount. This is one of the advantages of an electronic

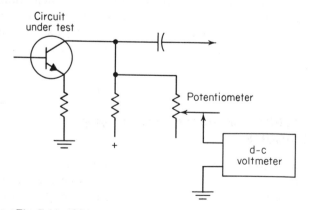

Fig. 7-11   Measuring internal resistance of circuits.

voltmeter. In some cases, even an electronic voltmeter will cause circuit loading.

Determining Voltage Sensitivity

The fact that a circuit is voltage-sensitive can be determined easily using the test setup of Fig. 7-12. This test setup could be applied to any circuit. (The AGC line of a receiver is a typical voltage-sensitive circuit.)

1. Measure the voltage directly as shown in Fig. 7-12(a).
2. Insert the series resistor R and measure the voltage at the same point in the circuit as shown in Fig. 7-12(b). The value of resistor R should be equal to the input resistance of the meter.
3. The second voltage indication (with resistor R in series) should be ap-

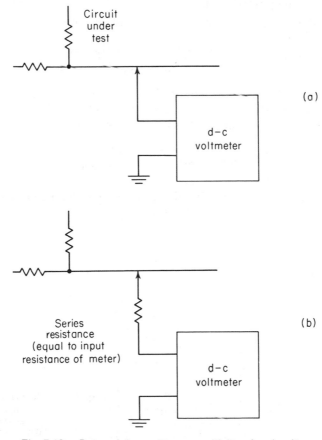

Fig. 7-12   Determining voltage sensitivity of a circuit.

proximately one-half that obtained by direct measurement. If the second voltage indication is *much greater* than one-half, the circuit is voltage-sensitive.

Either of the following two methods (potentiometer and opposing-voltage methods) can be used to measure voltage-sensitive circuits.

Potentiometer Method

1. Connect the equipment as shown in Fig. 7-13.
2. Set the potentiometer to zero and measure the circuit voltage. This will be the initial voltage.
3. Increase the potentiometer resistance until the voltage is one-half the initial voltage (one-half that obtained in step 2).
4. Disconnect the potentiometer from the circuit. Measure the d-c resistance of the potentiometer.
5. Find the true circuit voltage with the following equation:

$$\text{true voltage} = \frac{\text{initial voltage} \times \text{potentiometer resistance}}{\text{meter input resistance}}$$

For example, assume that the initial circuit reading was 3 V, the potentiometer resistance needed to reduce the reading to 1.5 V was 140,000 ohms, and the meter input resistance was 60,000 ohms.

$$3 \times 140,000 = 420,000 \qquad \frac{420,000}{60,000} = 7 \text{ V}$$

Therefore, the true circuit voltage is 7 V, and the meter caused a 4-V drop.

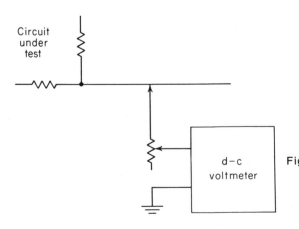

Fig. 7-13   Determining true voltage of a voltage-sensitive circuit (potentiometer method).

Opposing-Voltage Method

1. Connect the equipment as shown in Fig. 7-14(a). The value of series resistance R is not critical but should be near that of the meter input resistance.
2. Adjust the opposing-voltage source until the meter reads zero.
3. Disconnect the meter and opposing-voltage source.
4. Reconnect the meter to measure the opposing-voltage source as shown in Fig. 7-14(b).
5. The true circuit voltage is equal to the opposing-source voltage.

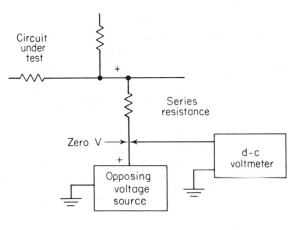

(a)

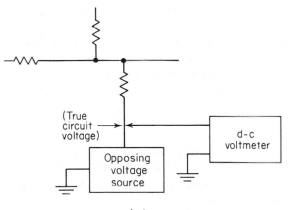

(b)

Fig. 7-14   Determining true voltage of a voltage-sensitive circuit (opposing voltage method).

7-8 MEASURING CIRCUIT Q

A resonant circuit has a Q, or quality, factor. A Q factor can also be applied to a complete circuit having resonant components, such as a tuned RF or IF stage of a receiver. The circuit Q is dependent upon the individual Q factors of the inductance and capacitance used in the circuit. If a circuit had pure inductance and capacitance, the response curve would be very sharp (or high Q). However, since any resonant circuit has some resistance, the Q factor is limited. A high Q is not always desired. The primary concern in a resonant circuit or stage is that a high Q will produce a sharp response curve, whereas a low Q will produce a broad response curve. Usually, a resonant circuit is measured at a point on either side of the resonant frequency where the signal amplitude is down to 0.707 of the peak resonant value. Other reference points can be used in special cases.

The Q of a circuit can be checked using a signal generator and a meter with an RF probe. An electronic meter will provide the least loading effect on the circuit and will therefore provide the most accurate indication. Figure 7-15(a) shows the test circuit for measuring Q in which the signal generator is connected directly to the input of a complete stage, and Fig. 7-15(b) shows the indirect method of connecting the signal generator to the input. When the stage or circuit has sufficient gain to provide a good reading on the meter with a nominal output from the generator, the indirect method is preferred. Any signal generator has some output impedance (such as a 50-ohm output resistor). When this resistance is connected directly to the tuned circuit, the Q is lowered, and the response becomes broader. (In some cases, the generator output impedance will seriously detune the circuit.) When it is not practical to use the indirect method, the generator output should be connected through an isolating resistor R. The value of R is not critical but should be near that of the stage ahead of the circuit being measured.

Fig. 7-15(c) shows the test circuit for a single component (such as an IF transformer). Since there is rarely enough gain in a transformer, the direct method (with an isolating resistor) must be used.

To measure circuit Q, the following steps are applicable:

1. Connect the equipment as shown in Fig. 7-15(a), (b), or (c), as the situation requires.
2. Tune the signal generator to the circuit resonant frequency. Operate the generator to produce an unmodulated output.
3. Tune the signal generator for maximum reading on the meter. Note the generator frequency.
4. Tune the signal generator below resonance until the meter reading is 0.707 of the maximum reading. Note the generator frequency. (To

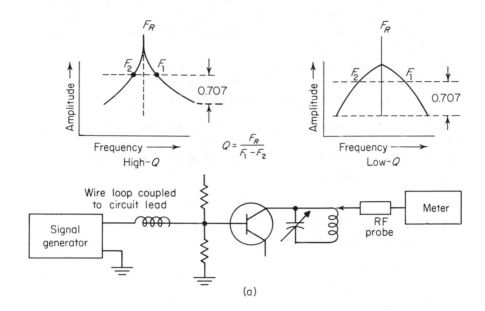

$$Q = \frac{F_R}{F_1 - F_2}$$

High-Q Low-Q

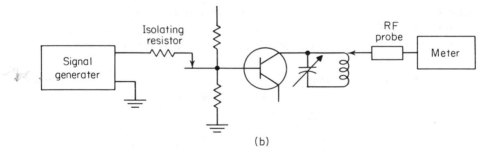

(a)

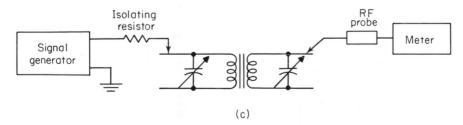

(b)

(c)

Fig. 7-15   Measuring circuit and component Q.

make the calculation more convenient, adjust the signal-generator output so that the meter reading is some even value such as 1 V or 10 V after the generator is tuned for maximum. This will make it easy to find the 0.707 mark.)

5. Tune the signal generator above resonance until the meter reading is 0.707 of the maximum reading. Note the generator frequency.

6. Calculate the circuit Q using the equation shown in Fig. 7-15.

For example, assume that the maximum meter indication occurred at 455 kHz (F_R) the below-resonance indication was at 453 kHz (F_2) and the above-resonance indication was at 458 kHz (F_1). $458 - 453 = 5$; $^{455}/_5$ $= 91$. 91 is the circuit Q value.

7-9 MEASURING STATIC VOLTAGES OF INTEGRATED CIRCUITS

The static (direct-current) voltages of an integrated circuit are measured in essentially the same manner as those of conventional transistor circuits, with one major exception. Most ICs require connection to *both a positive and negative* power source. A few ICs can be operated from a single power-supply source. Many ICs require equal power-supply voltages (such as $+9$ V and -9 V). However, this is not the case with the example circuit of Fig. 7-16, which requires a $+9$ V at pin 8 and a -4.2 V at pin 4.

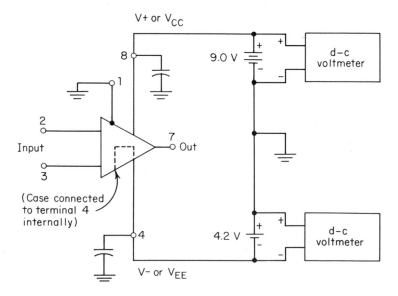

Fig. 7-16   Measuring static voltages on integrated circuits.

Unlike most transistor circuits, where it is common to label one power-supply lead positive and the other negative without specifying which (if either) is common or ground, *it is necessary that all IC power-supply voltages be referenced to a common or ground.*

Manufacturers do not agree on power-supply labeling for ICs. For example, the IC of Fig. 7-16 uses V + to indicate the positive voltage and V − to indicate the negative voltage. Another manufacturer might use the symbols V_{EE} and V_{CC} to represent negative and positive, respectively. As a result the IC data sheet should be studied carefully before measuring the power-source voltages.

No matter what labeling is used, the IC will usually require two power sources with the positive lead of one and the negative lead of the other tied to ground. Each voltage must be measured separately, as shown in Fig. 7-16. .

Note that the IC case (such as to a TO-5) of the Fig. 7-16 circuit is connected to pin 4. This is typical for most ICs. Therefore, the case will be below ground (or "hot") by 4.2 V.

7-10 MEASURING VOLTAGES IN CIRCUITS

It is possible to locate many defects in electronic circuits by measuring and analyzing voltages at the elements of active devices (grid, cathode, and plate of vacuum tubes; base, emitter, and collector of transistors). This can be done with the circuit operating and without disconnecting any parts. Once located, the defective part can be disconnected and tested or substituted, whichever is most convenient.

Vacuum-tube circuits can be analyzed with a simple VOM or electronic meter. The normal relationships of vacuum-tube elements are generally fixed. For example, the plate is positive, the cathode is at ground or positive, and the grid is (usually) negative.

Transistor circuits are best analyzed with an electronic voltmeter or with a very sensitive VOM. A number of manufacturers produce VOMs designed specifically for transistor troubleshooting. The meters of Figs. 3-1 and 3-12 are typical. These VOMs have very low voltage scales (0 to 0.3 V; 0 to 0.5 V) to measure the differences that often exist between elements of a transistor (especially the small voltage difference between emitter and base). Such VOMs also have a very low voltage drop in the current ranges.

Analyzing Transistor Voltages

Figure 7-17 shows the basic connections for both PNP and NPN transistor circuits. The coupling or bypass capacitors have been omitted to simplify the explanation. The purpose of Fig. 7-17 is to establish "normal" tran-

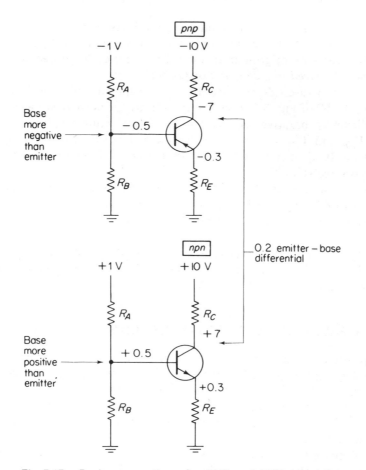

Fig. 7-17   Basic connections for PNP and NPN transistor circuits (with normal voltage relationships). (John D. Lenk, Handbook of Basic Electronic Troubleshooting, © 1977, p. 152. Courtesy Prentice-Hall.)

sistor voltages. With a "normal" pattern established, it is relatively simple to find an "abnormal" condition.

    In practically all transistor circuits, the emitter-base junction must be forward-biased to get current flow through a transistor. In a PNP transistor, this means that the base must be made more negative (or less positive) than the emitter. Under these conditions, the emitter-base junction will draw current and cause a heavy current to flow from the collector to the emitter. In an NPN transistor, the base must be made more positive (or less negative) than the emitter in order for current to flow from emitter to collector.

The following general rules can be helpful in a practical analysis of transistor voltages:

1. The middle letter in PNP or NPN always applies to the *base*.
2. The first two letters in PNP or NPN refer to the *relative* bias polarities of the *emitter* with respect to either the base or collector. For example, the letters PN (in PNP) indicate that the emitter is positive with respect to both the base and collector. The letters NP (NPN) indicate that the emitter is negative with respect to both the base and collector.
3. The collector-base junction is always reverse-biased.
4. The emitter-base junction is always forward-biased. An exception is a class C amplifier (used in RF circuits).
5. A *base input* voltage that opposes or decreases the forward bias also decreases the emitter and collector currents.
6. A *base* input voltage that aids or increases the forward bias also increases the emitter and collector currents.
7. The d-c *electron flow* is always against the direction of the arrow on the emitter.
8. If electron flow is into the emitter, electron flow will be out from the collector.
9. If electron flow is out from the emitter, electron flow will be into the collector.

Using these rules, "normal" transistor voltages can be summed up this way:

1. For a PNP, the base is negative, the emitter is not quite as negative, and the collector is far more negative.
2. For an NPN, the base is positive, the emitter is not quite as positive, and the collector is far more positive.

Measurement of Transistor Voltages

There are two schools of thought on how to measure transistor voltages. Some technicians prefer to measure transistor voltages from element to element and note the difference in voltage. For example, in the circuits of Fig. 7-17, an 0.2-V differential would exist between emitter and base. Likewise, a 9.5-V differential would exist between base and collector. The element-to-element method of measuring transistor voltages will quickly establish forward and reverse bias.

The most common method of measuring transistor voltages is to measure from a common or ground to the element. Service literature generally specifies transistor voltages in this way. For example, all the voltages for the PNP of Fig. 7-17 are negative with respect to ground. (The positive test lead of the meter must be connected to ground, and the negative test lead is connected to the elements in turn.)

This method of labeling transistor voltages is sometimes confusing to those not familiar with transistors, since it appears to break the rules. (In a PNP, both the emitter and collector should be positive, yet all of the elements are negative.) However, the rules still apply.

In the case of the PNP of Fig. 7-17, the emitter is at -0.3V, whereas the base is at -0.5V. The base is *more negative* than the emitter. Therefore, the emitter is *positive with respect to the base*, and the base-emitter junction is forward-biased (normal).

On the other hand, the base is at -0.5V, whereas the collector is at -10V. The base is *less negative* than the collector. Therefore, the *base* is *positive with respect to the collector*, and the base-collector junction is reverse-biased (normal).

Troubleshooting with Transistor Voltages

This section presents an example of how voltages measured at the elements of a transistor can be used to analyze failure in solid-state circuits.

Assume that an NPN transistor is measured and that the voltages found are similar to those shown in Fig. 7-18. Except in one case, these voltages indicate a defect. It is obvious that the transistor is not forward-biased because the base is less positive than the emitter (reverse bias for an NPN). The only circuit where this might be normal is one that requires a large *trigger voltage* or pulse (positive in this case) to turn it on.

The first troubleshooting clue in Fig. 7-18 is that the collector voltage is almost as large as the collector source (at R_C). This means that very little current is flowing through R_C in the collector-emitter circuit. The transistor could be defective. However, the trouble is more likely caused by a problem in bias. The emitter voltage depends mostly on the current through R_E. Therefore, unless the value of R_E has changed substantially (this would be unusual), the problem is one of incorrect bias on the base.

The next step in this case is to measure the bias-source voltage at R_A. If the bias-source voltage is (as shown in Fig. 7-19) at 0.7 V instead of the required 2 V, the problem is obvious; the external bias voltage is incorrect. This condition will probably show up as a defect in the power supply and will appear as an incorrect voltage in other circuits.

If the bias source voltage is correct, as shown in Fig. 7-20, the cause of the trouble is probably a defective R_A or R_B, or a defect in the transistor.

The next step is to remove all voltage from the equipment and measure the resistance of R_A and R_B. If either value is incorrect, it is reasonable

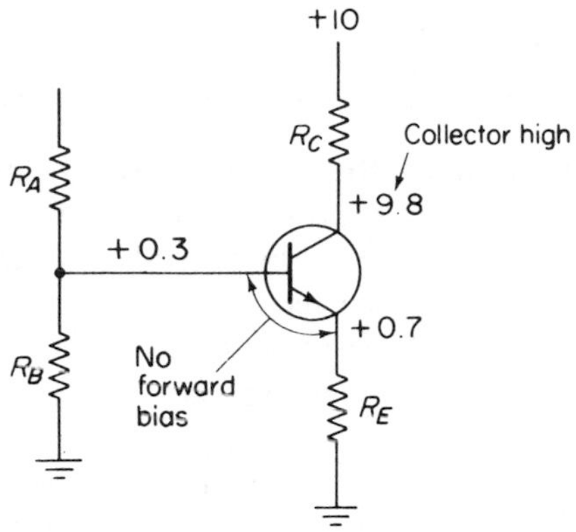

Fig. 7-18 NPN transistor circuit with abnormal voltages (emitter base not forward biased, collector voltage high). (John D. Lenk, Handbook of Basic Electronic Troubleshooting,© 1977, p. 154. Courtesy Prentice-Hall.)

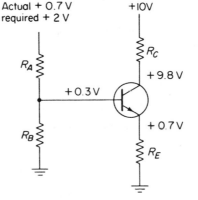

Fig. 7-19 NPN transistor circuit with abnormal voltages (fault traced to incorrect bias source, bias voltage low). (John D. Lenk, Handbook of Basic Electronic Troubleshooting, © 1977, p. 155. Courtesy Prentice-Hall.)

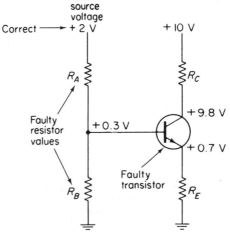

Fig. 7-20 NPN transistor circuit with abnormal voltages (fault traced to bias resistors or transistor). (John D. Lenk, Handbook of Basic Electronic Troubleshooting, © 1977, p. 155. Courtesy Prentice-Hall.)

to check the value of R_E. However, it is more likely that the transistor is defective. This can be established by test and/or replacement.

Practical In-Circuit Resistance Measurements Do not attempt to measure resistance values in transistor circuits with the resistors still connected (as discussed in Sec. 3-8). Although this practice may be correct for vacuum-tube circuits, it is incorrect for transistor circuits. One reason is that the voltage produced by the ohmmeter battery could damage some transistors.

Even if the voltages are not dangerous, the chance for an error is greater with a transistor circuit because the transistor junctions will pass current in one direction. This can complete a circuit through other resistors and produce a series or parallel combination, thus making false indications. This can be prevented by *disconnecting one resistor lead* before making the resistance measurement.

For example, assume that an ohmmeter is connected across R_B (Figs. 7-18 to 7-20) with the negative battery terminal of the ohmmeter connected to ground, as shown in Fig. 7-21. Because R_E is also connected to ground, the negative battery terminal is connected to the end of R_E. Because the positive battery terminal is connected to the transistor base, the base-emitter junction is forward-biased and there is electron flow. In effect, R_E is now *in parallel* with R_B, and the ohmmeter reading is incorrect. This can be prevented by disconnecting either end of R_B before making the measurement. As an alternative, if a meter such as shown in Fig. 3-6 is available, it is possible to use the low-voltage-resistance scale for such measurements. The

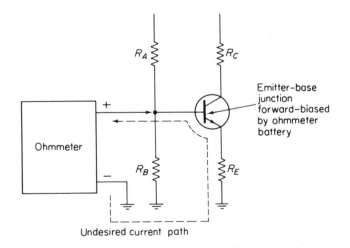

Fig. 7-21   Example of in-circuit resistance measurements showing undesired current path through forward-biased transistor junction. (John D. Lenk, Handbook of Basic Electronic Troubleshooting, © 1977, p. 156. Courtesy Prentice-Hall.)

maximum meter voltage, on the low-voltage scale, is below the threshold or turn-on voltage of most transistors.

Testing Transistors in Circuit (Forward-Bias Method)

Germanium transistors normally have a *voltage differential* of 0.2 to 0.4 V between emitter and base. Silicon transistors usually have a voltage differential of 0.4 to 0.8 V. The polarities of voltages at the emitter and base depend upon the type of transistor (NPN or PNP).

The voltage differential between emitter and base acts as a forward bias for the transistor. That is, a sufficient differential or forward bias will turn the transistor on, resulting in a corresponding amount of emitter-collector flow. Removal of the voltage differential or an insufficient differential will produce the opposite results. That is, the transistor is cut off (no emitter-collector flow or very little flow).

These forward-bias characteristics can be used to troubleshoot circuits without removing the transistor and without using an in-circuit tester. The following paragraphs describe two methods of testing transistors in circuit using a voltmeter: one by removing the forward bias and the other by introducing a forward bias.

Removal of Forward Bias Figure 7-22 shows the test connections for an in-circuit transistor test by removal of forward bias. The procedure is simple. First, measure the emitter-collector differential voltage under normal circuit conditions. Then, short the emitter-base junction and note any change in emitter-collector differential. If the transistor is operating, the removal of forward bias causes the emitter-collector current flow to stop, and the emitter-collector voltage differential increases. That is, the collector voltage rises to or near the power-supply value.

For example, assume that the power-supply voltage is 10 V and that the differential between the collector and emitter is 5 V when the transistor is operating normally (no short between emitter and base). When the emitter-base junction is shorted, the collector-emitter differential should rise to about 10 V (probably somewhere between 9 and 10 V).

Application of Forward Bias Figure 7-23 shows the test connection for an in-circuit transitor test by the application of forward bias. The procedure is equally simple. First, measure the emitter-collector differential under normal circuit conditions. As an alternative, measure the voltage across R_E, as shown in Fig. 7-23.

Next, connect a 10-kilohm resistor between the collector and base, as shown, and note any change in emitter-collector differential (or any change in voltage across R_E). If the transistor is operating, the application of for-

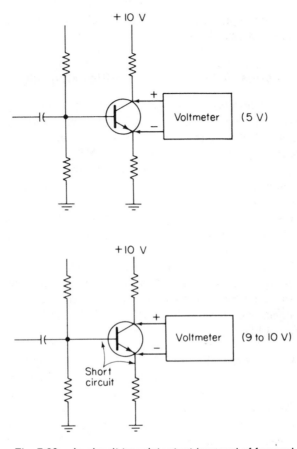

ward bias will cause the emitter-collector current flow to start (or increase), and the emitter-collector voltage differential will decrease, or the voltage across R_E *will increase.*

Go/No-Go Test Characteristics The test methods shown in Figs. 7-22 and 7-23 show that the transistor is operating on a go/no-go basis. This is usually sufficient for most d-c and low-frequency a-c applications. However, the tests do not show transistor gain or leakage. Also, the tests do not establish operation of the transistor at high frequencies or show how much delay is introduced by the transistor.

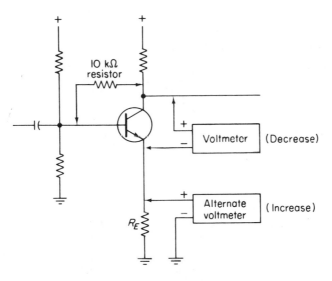

Fig. 7-23. In-circuit transistor test (application of forward bias). (John D. Lenk, Handbook of Basic Electronic Troubleshooting, © 1977, p. 158. Courtesy Prentice-Hall.)

7-11 MEASURING RESONANT FREQUENCY OF LC CIRCUITS

A meter can be used in conjunction with a signal generator (audio or RF) to find the resonant frequency of either series or parallel LC circuits (such as tank circuits, filters, etc.). The generator must be capable of producing a signal at the resonant frequency of the circuit, and the meter must be capable of measuring the frequency. If the resonant frequency is beyond the normal range of the meter, an RF probe must be used. Use the following procedure to measure LC-circuit resonant frequency:

1. Connect the equipment as shown in Fig. 7-24. Use the connections of Fig. 7-24(a) for a parallel resonant circuit and the connections of Fig. 7-24(b) for a series resonant circuit.

2. Adjust the generator output until a convenient midscale indication is obtained on the meter. Use an unmodulated signal output from the generator.

3. Starting at a frequency well below the lowest possible frequency of the circuit under test, slowly increase the generator output frequency. If there is no way to judge the approximate resonant frequency, use the lowest generator frequency.

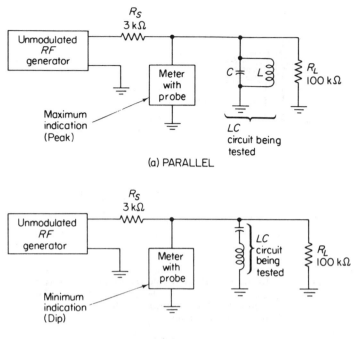

(a) PARALLEL

(b) SERIES

Fig. 7-24   Measuring resonant frequency of LC circuits. (John D. Lenk, Handbook of Basic Electronic Troubleshooting, © 1977, p. 216. Courtesy Prentice-Hall.)

4. If the circuit being tested is parallel-resonant, watch the meter for a maximum or peak indication.

5. If the circuit being tested is series-resonant, watch the meter for a minimum or dip indication.

6. The resonant frequency of the circuit under test is the one at which there is a maximum (for parallel) or minimum (for series) indication on the meter.

7. There may be peak or dip indications at harmonics of the resonant frequency. Therefore, the test is most efficient when the approximate resonant frequency is known.

8. The value of load resistor R_L is not critical. The load is shunted across the LC circuit to flatten or broaden the resonant response. Thus, the voltage maximum or minimum is approached more slowly. A suitable trial value for R_L is 100,000 ohms. A lower value of R_L will sharpen the resonant response, and a higher value will flatten the curve.

Determining Capacitance in LC Circuits

When the inductance value of an LC circuit is known, it is possible to find the capacitance value using a meter and a signal generator.

1. Connect the equipment as shown in Fig. 7-25. Use an inductance value such as 10 mH, 100 mH, or some even number to simplify the calculation.

2. Adjust the generator output until a convenient midscale indication is obtained on the meter. Use an unmodulated signal output from the generator.

3. Starting at a frequency well below the lowest possible resonant frequency of the inductance-capacitance combination under test, slowly increase the generator frequency. If there is no way to judge the approximate resonant frequency, use the lowest generator frequency.

4. Watch the meter for a maximum or peak indication. Note the audio frequency at which the peak indication occurs. This is the resonant frequency of the circuit.

5. Using this resonant frequency and the known inductance value, calculate the unknown capacitance using Fig. 7-25.

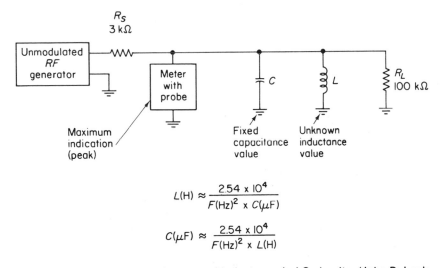

$$L(H) \approx \frac{2.54 \times 10^4}{F(Hz)^2 \times C(\mu F)}$$

$$C(\mu F) \approx \frac{2.54 \times 10^4}{F(Hz)^2 \times L(H)}$$

Fig. 7-25 Measuring capacitance and inductance in LC circuits. (John D. Lenk, Handbook of Basic Electronic Troubleshooting, © 1977, p. 217. Courtesy Prentice-Hall.)

Determining Inductance in LC Circuits

When the capacitance value of an LC circuit is known, it is possible to find the inductance value using a meter and a signal generator. The procedure is the same as for finding capacitance, except that a known capacitance value is inserted in the test circuit instead of a known inductance. Use an even capacitance value such as 10 μF or 100 μF in the circuit of Fig. 7-25. When the resonant frequency is found (a peak indication on the meter), calculate the unknown inductance value using Fig. 7-25.

Measuring Self-Resonance and Distributed Capacitance of a Coil

There is distributed capacitance in any coil, which can combine with coil inductance to form a resonant circuit. Although the self-resonant frequency may be high in relation to the operating frequency at which the coil is used, it may be near a harmonic of that operating frequency. This limits the usefulness of the coil in an LC circuit. Some coils, particularly RF chokes used in transmitters, may have more than one self-resonant frequency.

The circuit for measuring self-resonance and distributed capacitance of a coil is shown in Fig. 7-26. To use the circuit, adjust the unmodulated RF generator output amplitude for a convenient indication on the meter. Tune the generator over its entire frequency range, starting at the lowest frequency. Watch for either peak or dip indications on the meter. Either a peak or dip indicates that the inductance is at a self-resonant point. The generator output frequency at this point is the self-resonant frequency (or a harmonic of the frequency).

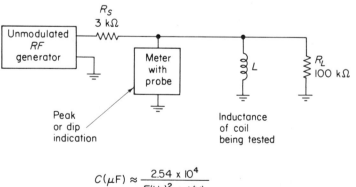

$$C(\mu F) \approx \frac{2.54 \times 10^4}{F(Hz)^2 \times L(H)}$$

Fig. 7-26   Measuring self-resonance and distributed capacitance of a coil. (John D. Lenk, Handbook of Basic Electronic Troubleshooting, © 1977, p. 217. Courtesy Prentice-Hall.)

Make certain that peak or dip indications are not the result of changes in generator output level. Cover the entire frequency range of the generator or at least from the lowest up to the third harmonic of the highest-frequency range involved in circuit design or operation.

Once the resonant frequency (or frequencies) has been found, calculate the distributed capacitance using the equation given in Fig. 7-26.

8

Servicing Specific Circuits with Meters

If a technician was limited to one item of test equipment for servicing, a meter would be the logical choice. While certain equipment may require an oscilloscope for a complete test, most servicing procedures can be handled with only a meter.

Previous chapters describe how the meter can be used effectively to test specific components and general circuit conditions. This chapter describes procedures for test and alignment of specific circuits such as receivers, transmitters, and amplifiers.

To be most effective in equipment service, a meter should be equipped with an RF probe (Chapter 2). Then, in addition to measuring voltage, current, and resistance in the equipment circuits, the meter can be used to *trace signals* through all stages of such units as receivers, transmitters, and amplifiers. When the signals are found to be absent or abnormal in a particular stage, trouble is localized to that stage. The defective part can then be isolated by means of voltage, current, or resistance checks (Chapters 6 and 7).

8-1 RECEIVER ALIGNMENT

It is possible to align and test receiver circuits using only a meter and a signal generator. Both AM and FM receivers require alignment of the IF and RF amplifiers. An FM receiver also requires alignment of the detector stage (discriminator or ratio detector). If the receiver includes an AVC (automatic volume control) or AGC (automatic gain control) circuit, the AVC or AGC circuit must be disabled (by means of a fixed bias voltage, usually on the order of a few volts).

FM Detector Alignment

The circuits for FM detector alignment are shown in Fig. 8-1. Note that both ratio-detector and discriminator circuits are given. In both cases, the signal to be injected is an unmodulated RF signal tuned to the intermediate frequency (IF) of the receiver. The signal can be injected at the input of the first IF stage, or at the primary winding of the detector transformer (which is also the final or output transformer of the IF stages).

With either circuit, a d-c voltmeter is required to monitor the detector output. However, the points at which the meter is connected into the circuit are different for each type of detector, as shown in Fig. 8-1.

If necessary, disable the AVC (or AGC) line by applying a fixed d-c voltage between the line and ground, as shown in Fig. 8-2. In the absence of specific values, always use the same polarity as the normal AVC or AGC voltage, and a value that is higher than the average AVC voltage (about twice the average value). For example, if the AVC line usually varies between 0 and -1 V, use a value of -2 V.

Adjust the signal generator frequency to the intermediate frequency (typically 10.7 MHz for household FM receivers). Use an *unmodulated* output from the signal generator.

Adjust the *secondary* winding of the transformer (either the capacitor or a tuning slug within the winding) for a *zero reading* (or a dip) on the meter. Adjust the transformer slightly each way, and make sure that the meter moves smoothly above or below the exact zero mark (or the minimum dip point). A meter with a zero-center scale, as described in Sec. 1-2 and shown in Fig. 1-9, is most helpful when adjusting FM detectors.

Adjust the signal generator to some point below the intermediate frequency (to 10.625 MHz for an FM detector with a 10.7-MHz IF). Note the meter reading.

Adjust the signal generator to some point above the intermediate frequency *exactly equal* to the amount set below the intermediate frequency. For example, if the generator is set to 0.075 below the IF, set the generator to 0.075 above the IF (or to 10.625 and 10.775 MHz, respectively for an 10.7-MHz IF).

The meter reading should be *approximately the same* on both sides of the IF, except that the polarity is reversed. For example, if the meter reading is 7 scale divisions below zero and 7 scale divisions above zero (on a zero-center meter), the FM detector is balanced. If an FM detector cannot be

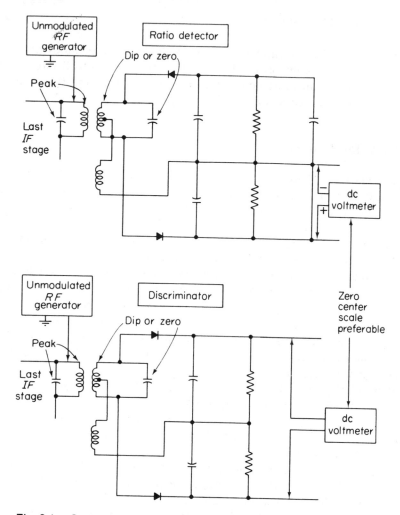

Fig. 8-1   Connections for FM detector alignment. (John D. Lenk, Handbook of Basic Electronic Troubleshooting, © 1977, p. 202. Courtesy Prentice-Hall.)

balanced, the fault is usually a *serious mismatch* in diodes or other components.

Return the generator to the IF (10.7 MHz), and adjust the *primary winding* of the transformer for *maximum* or *peak* reading on the meter. This sets the primary winding at the correct resonant frequency.

AM and FM Alignment

The alignment procedures for the IF amplifier stages of an AM receiver are essentially the same as those used for an FM receiver. However, the meter must be connected at different points in the corresponding detector, as

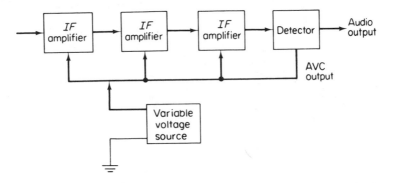

Fig. 8-2   Connections for disabling AVC line. (John D. Lenk, Handbook of Basic Electronic Troubleshooting, © 1977, p. 203.
Courtesy Prentice-Hall.)

shown in Fig. 8-3. In either case, the meter is set to measure direct current
and the RF probe is not used.

In those cases in which the IF stages are being tested without a detector (such as during design or after extensive troubleshooting), an RF probe
is required. As shown in Fig. 8-3, the RF probe is connected to the secondary winding of the final IF output transformer.

If necessary, disable the AVC or AGC line as described for FM detector alignment. Then, set the meter to measure direct current, and connect
the meter to the appropriate test point (with or without an RF probe, as applicable). Adjust the generator frequency to the receiver IF (typically 10.7
MHz for FM and 455 kHz for AM household receivers). Use an *unmodulated* RF signal.

Adjust the windings of the IF transformers (capacitor or tuning slug)
in turn, starting with the last stage and working toward the first stage. Adjust each winding for *maximum* reading. Repeat the procedure to make sure
that there is no interaction between adjustments (usually there may be some
small interaction).

AM and FM RF Amplifier Alignment

The alignment procedures for the RF stages (RF amplifiers, local oscillator,
mixer-converter) of an AM receiver are essentially the same as those used
for an FM receiver. Again, it is a matter of connecting the meter to the appropriate test point. The same test points used for IF alignment can be used
for aligning the RF stages, as shown in Fig. 8-4. However, if an individual
RF stage is to be aligned, the meter must be connected to the secondary winding of the RF stage output transformer through an RF probe.

If necessary, disable the AVC or AGC line as described for FM detector alignment. Set the meter to measure direct current and connect the meter
to the appropriate test point (with or without an RF probe, as applicable).

Adjust the generator frequency to some point near the high end of the

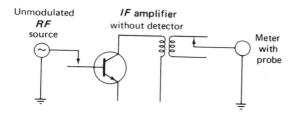

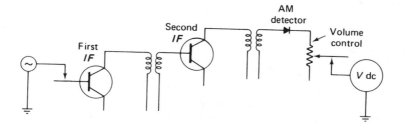

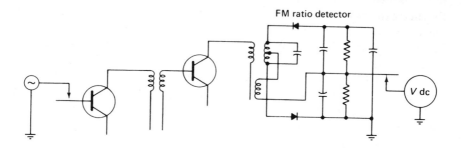

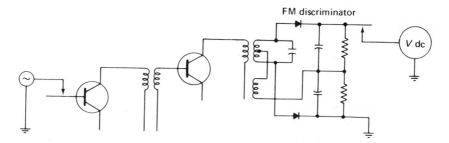

Fig. 8-3   Connections for IF alignment of AM and FM receivers. (John D. Lenk, Handbook of Basic Electronic Troubleshooting, © 1977, p. 205. Courtesy Prentice-Hall.)

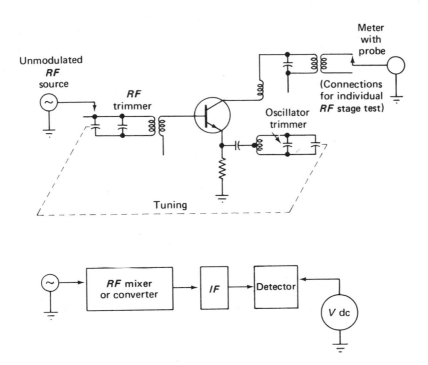

Fig. 8-4 Connections for FM amplifier and local-oscillator alignment of AM and FM receivers. (John D. Lenk, Handbook of Basic Electronic Troubleshooting, © 1977, p. 206. Courtesy Prentice-Hall.)

receiver operating frequency (typically 107 MHz for an FM broadcast receiver and 1400 kHz for an AM broadcast receiver). Use an unmodulated output from the signal generator.

Adjust the trimmer of the RF stage for *maximum* reading on the meter.

Adjust the generator frequency to the low end of the receiver operating frequency (typically 90 MHz for FM and 600 kHz for AM).

Adjust the trimmer of the oscillator stage for *maximum* reading on the meter.

Repeat the procedure to make sure that the resonant circuit tracks across the entire tuning range of the receiver.

8-2 TRACING SIGNALS IN RECEIVER CIRCUITS

Signals can be traced through all sections of both FM and AM receivers. An RF probe is required for signal tracing in the RF and IF stages. The audio stages of a receiver are essentially amplifiers. Tracing signals in amplifier circuits is discussed in Secs. 8-9 through 8-12. The following procedures cover signal tracing in receiver circuits from the front end to the audio stages:

1. Connect the equipment as shown in Fig. 8-5.

2. Use the meter RF probe at all test points except the detector. Set the meter to measure dc and connect it to the appropriate detector test point.

3. Place the signal generator in operation and adjust the generator frequency to some point near the center of the receiver operating frequency. Use an unmodulated output from the signal generator.

4. Connect the meter to each of the test points in turn, starting with the input or antenna and working toward the detector.

5. Usually, the output of a stage will produce a higher-voltage indication than the input. However, this is not always the case. Always consult the receiver service data.

6. If there is some doubt that the voltage indication is being obtained from RF signals (through the RF probe) rather than from circuit voltages, switch the generator on and off and check the meter reading.

7. Trouble is isolated by noticing where the trouble first begins. For example, assume that a receiver is completely dead, yet a signal is traced satisfactorily from the antenna to the output of the detector. The trouble is obviously beyond this point or in the amplifier stages.

8-3 MEASURING RECEIVER SELECTIVITY

In radio receivers, selectivity is the ability to pick out a signal on one specific frequency. Although any unit of measurement can be used to express selectivity, decibels are most often used since they express a ratio. Selectivity is

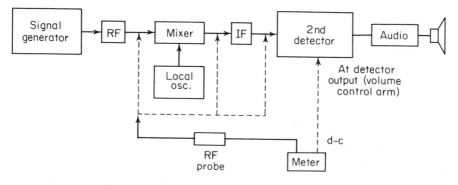

Fig. 8-5   Signal tracing receiver circuits.

the ratio of the desired signal to the nearby undesired signals. The following is the procedure for measuring receiver selectivity:

1. Connect the equipment as shown in Fig. 8-6.

2. Set the meter to measure ac or "output."

3. Place the generator in operation and adjust the generator frequency to some point near the low end of the receiver's operating range (600 kHz for broadcast AM, 90 Mhz for broadcast FM, etc.). Use a modulated output from the signal generator.

4. Tune the receiver to the generator frequency. Tune for a peak indication on the meter. Note the meter indication in decibels. The actual dB value is not important; only the relative change in dB value.

5. Tune the generator to a point 10 kHz above the receiver frequency. Do not change the receiver frequency. Note the new decibel reading.

6. Tune the generator to a point 10 kHz below the receiver frequency and note the new decibel reading.

7. As shown in the following equation, the selectivity is the ratio of on-frequency to 10 kHz off-frequency readings in decibels. Repeat the tests at various points across the entire turning range of the receiver.

$$\text{selectivity} = \frac{\text{on-frequency output reading in decibels}}{\text{off-frequency output reading in decibels}}$$

8. The output indication will drop off sharply on a receiver with good selectivity. Also, the dropoff should be approximately the same on either side of the on-frequency. If not, selectivity is poor.

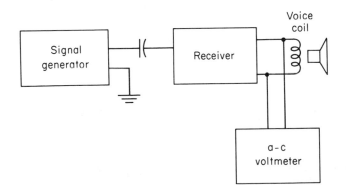

Fig. 8-6   Measuring receiver selectivity, sensitivity, and image rejection ratio.

8-4 MEASURING RECEIVER SENSITIVITY

In radio receivers, sensitivity is the measure of output for a given input or the required input for a given output. Although standards vary, most communications receivers and many broadcast receivers specify a certain number of microvolts input for each 0.5 W of output. To make accurate sensitivity tests, it is therefore necessary to measure the generator output in microvolts. The more expensive laboratory-type generators have a built-in output meter. If such a generator is not available, it is possible to measure the generator output with a sensitive electronic voltmeter. The following procedure outlines the steps used in measuring receiver sensitivity:

1. Connect the equipment as shown in Fig. 8-6.
2. Set the meter to measure ac or "output."
3. Place the generator in operation and adjust the generator frequency to some point near the center of the receiver operating range. Use a modulated output from the signal generator. For a standard test, use 30 percent modulation at 400 Hz.
4. Tune the receiver to the generator frequency. Tune for a peak indication on the meter.
5. Calculate the power across the loudspeaker voice coil using the following equation:

$$\text{power across resistor} = \frac{(\text{voltage across voice coil})^2}{\text{voice coil impedance}}$$

6. The procedure for measuring impedance of the loudspeaker voice coil is described in Sec. 7-2. An alternative impedance measuring procedure is discussed in Sec. 8-10.
7. Starting from zero, increase the signal-generator output voltage until the power across the voice coil is 0.5 W. For example, assume that the voice coil was 8 ohms. When the voltage across the voice coil was 2V, the power output would be 0.5 W.

$$\text{power} = \frac{2^2}{8} = 0.5 \text{ W}$$

8. With the power output set at 0.5 W, read the signal-generator output in microvolts. This is the receiver sensitivity.

8-5 MEASURING RECEIVER IMAGE-REJECTION RATIO

The image frequency of a radio receiver is equal to the indicated tuning-dial frequency plus or minus *twice* the intermediate frequency. For example, assume an intermediate frequency of 455 kHz and a tuning-dial frequency of 600 kHz. The receiver's local oscillator would be at 1055 kHz and would mix with the 600-kHz signal to produce an IF of 455 kHz. Another signal at 1510 kHz (600 kHz plus twice 455, or 910 kHz) would also combine with the 1055-kHz oscillator signal to produce a 455-kHz IF. This image signal at 1510 kHz could pass the receiver's preselector circuits (if they are improperly tuned or defective). Image-rejection ratio is a measure of the receiver's ability to reject the image signal (1510 kHz in this case) while passing the desired signal (600 kHz). The higher the image-rejection ratio, the better the preselection. These are the steps to use when measuring image-rejection ratios:

1. Connect the equipment as shown in Fig. 8-6 (the same as for selectivity).
2. Set the meter to measure ac or "output."
3. Place the generator in operation and adjust the generator frequency to some point near the low end of the receiver operating range. Use a modulated output from the signal generator.
4. Tune the receiver to the generator frequency. Tune for a peak indication on the meter. Note the meter reading.
5. Tune the generator to the image frequency. Note the meter reading.
6. The image-rejection ratio of the receiver is the ratio of the two meter readings.

8-6 MEASURING OVERALL AUDIO RESPONSE IN RECEIVERS

The audio section of a receiver can be treated as an amplifier and tested as described in Sec. 8-9 through 8-12. It is also possible to check the overall audio response of a receiver, using a signal generator and an external audio oscillator. This type of test not only determines the frequency response of the audio section but also shows the condition of the signal circuits (bandwidth, alignment, etc.).

1. Connect the equipment as shown in Fig. 8-7.
2. Set the meter to measure ac or "output."
3. Place the signal generator and audio generator in operation. Adjust the signal generator to some point near the center of the receiver operating range. Modulate the signal generator with the external audio generator.
4. Tune the receiver to the generator frequency. Tune for a peak indication on the meter.
5. Vary the audio-generator frequency throughout the audio range of the receiver. Note the meter reading at each audio frequency. If desired,

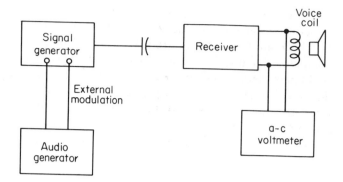

Fig. 8-7   Measuring overall audio response in receiver.

draw an audio response curve as described in Sec. 8-10. Make certain to observe the precautions of Sec. 8-10 regarding the setting of tone and equalizer controls.

6. Repeat the test at various operating frequencies of the receiver. Retune the receiver for maximum output indication at each new signal-generator frequency. The audio response should remain the same over the entire operating range of the receiver. Check this against the receiver service data.

8-7 MEASURING CHARACTERISTICS OF PORTABLE TRANSISTOR RECEIVERS

One of the problems in testing "transistor portables" is injecting the generator output into the circuit being tested. This can be done using an injection coil as shown in Fig. 8-8. The signal-generator output is connected to the coil, and the portable is placed near the coil. If necessary, the receiver and coil can be moved together or apart to provide the correct signal. The effects of hand capacitance can be minimized by moving the coil further from the portable; the signal can be increased by moving the coil closer to the portable.

Another problem in testing transistor-portable characteristics is measuring the signal at the loudspeaker voice coil. Often the loudspeaker is inaccessible. More frequently, the voltage across the loudspeaker is not sufficient to give a good indication on the meter. One way to overcome this problem is to measure the AGC-AVC line voltage. This voltage varies with the input signal. However, make sure which way the AGC-AVC voltage varies in response to a signal increase. In vacuum-tube receivers, the AGC-AVC voltage is usually negative and becomes more negative as the input signal increases. This is not always the case with transistor receiver circuits.

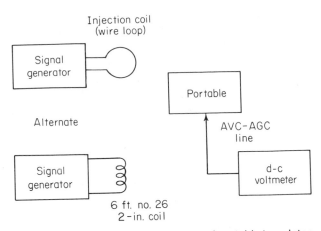

Fig. 8-8   Measuring characteristics of portable transistor receivers.

8-8 SERVICING TRANSMITTER CIRCUITS

It is possible to test and align all the transmitter RF stages using a meter with an RF probe. If an RF probe is not available, it is possible to use a circuit such as shown in Fig. 8-9.

The modulator stages of a transmitter are essentially audio amplifiers. Servicing amplifier circuits is discussed in Sec. 8-9 through 8-12. The following is the transmitter-circuit servicing procedure:

1. Connect the equipment as shown in Fig. 8-10.

2. Place the transmitter in operation as described in the instruction manual. Set the transmitter for an unmodulated carrier output.

3. In turn, connect the meter (through an RF probe or the special circuit of Fig. 8-9) to each stage of the transmitter. Start with the first stage (oscillator) and work toward the final (or output) stage.

4. A voltage indication should be obtained at each stage. Usually, the voltage indication will increase with each amplifier stage. Some stages may be frequency multipliers and provide no voltage amplification.

5. If a particular stage is to be tuned, adjust the tuning control for a maximum reading on the meter or as specified in the transmitter service data.

6. It should be noted that this tuning method will not guarantee that each stage is at the desired operating frequency. It is possible to get maximum readings on harmonics. However, most commercial transmitters are designed so that the tank circuit will tune on either side of the desired frequency but not to a harmonic (unless the circuit is seriously

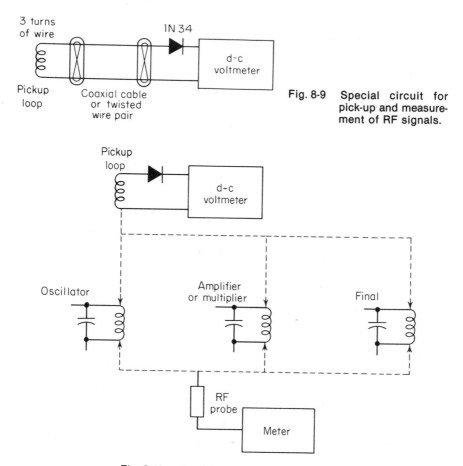

Fig. 8-9 Special circuit for pick-up and measurement of RF signals.

Fig. 8-10 Servicing transmitter circuits.

detuned). Therefore, the method should be satisfactory unless not recommended by the service data.

8-9 TRACING SIGNALS IN AMPLIFIER CIRCUITS

The major problem in using meters for service of amplifiers is the limited frequency range. It is possible to use almost any meter effectively with audio amplifiers (up to 15 or 20 kHz). However, the amplifiers used in industrial applications operate at much higher frequencies. Also, to be really effective, a meter should be capable of measuring signals up to about 500 kHz. This will ensure that any *harmonics* or *overtones* will be properly amplified and displayed, even though the manufacturers of the most advanced high-fidelity equipment claim bandpass characteristics no higher than 100 kHz.

If the frequency range of the amplifier circuit is beyond that of the meter or it is suspected that harmonics will be produced beyond the meter's capability, use an RF probe.

The procedure for signal tracing an amplifier is similar to that for a radio receiver. A sine wave is introduced into the input by means of an external generator. The amplitude of the input signal is measured on the meter. The meter probe or test lead is then moved to the input and output of each stage in turn, until the final output (usually at a loudspeaker or output transformer) is reached. The gain of each stage is measured as a *voltage* rather than as a dB value on the meter dB scales. As explained in Chapter 3, the meter dB scales show power ratios and are related to a given load value (usually 600 ohms). Since the input and output of each stage in an amplifier are rarely 600 ohms, a voltage ratio is more realistic. The following is the procedure for tracing signals in amplifier circuits:

1. Connect the equipment as shown in Fig. 8-11.
2. Set the meter to measure ac or "output," unless a probe is used.
3. Place the generator in operation. Unless otherwise specified by amplifier data, set the generator output frequency to 1000 Hz. Set the generator output level to the value recommended in the amplifier service data. Do not overload the amplifier.
4. Measure the output of the generator as it appears across the input of the first stage.
5. Move the meter probe or test lead to the output of the first amplifier stage. Measure and record the voltage.
6. Repeat step 5 for each stage of the amplifier from input to output. If

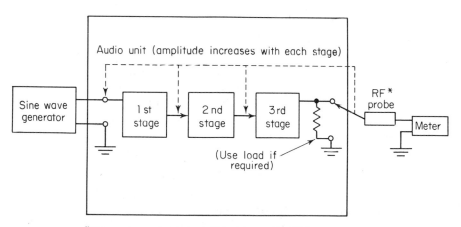

*RF probe not required if test is made within a-c frequency range of meter

Fig. 8-11   Signal-tracing amplifier circuits.

voltage at a stage output is lower than the input (or absent), it is likely that the particular stage is defective.

7. One factor often overlooked in testing amplifiers is setting the amplifier *amplitude, tone,* and *equalizer* controls to their normal operating point or to some particular point specified in the manufacturer's test data.

8. If it is desired to convert the voltage gain of one stage, a group of stages, or the complete amplifier into decibels, use this equation:

$$\text{dB gain} = 20 \log \frac{\text{output voltage}}{\text{input voltage}}$$

Or use the simplified dB conversion chart of Table 3-2.

9. The gain (or loss) of an amplifier component, such as a transformer or filter, can also be measured as described in steps 5 through 8.

8-10 MEASURING AMPLIFIER CHARACTERISTICS

The following sections describe the procedures for measuring various amplifier characteristics using a meter.

Amplifier Frequency Response

The frequency response of an audio amplifier can be measured with an audio-signal generator and a meter. The signal generator is tuned to various frequencies, and the resultant output response is measured at each frequency. The results are then plotted in the form of a graph or response curve, as shown in Fig. 8-12.

The basic procedure for measurement of frequency response is to apply a *constant-amplitude signal* while monitoring the amplifier output. The input signal is varied in frequency (but not in amplitude) across the entire operating range of the amplifier. Any well-designed audio amplifier should have a constant response from about 20 Hz to 20 kHz. With direct-coupled amplifiers, the response is usually extended from a few hertz (or possibly from direct current) up to 100 kHz (and higher). The voltage output at various frequencies across the range is plotted on a graph as follows:

1. Connect the equipment, as shown in Fig. 8-12.

2. Initially, set the generator frequency to the low end of the range. Then set the generator amplitude to the desired input level.

3. In the absence of a realistic test input voltage, set the generator amplitude to an arbitrary value. A simple method of finding a satisfactory

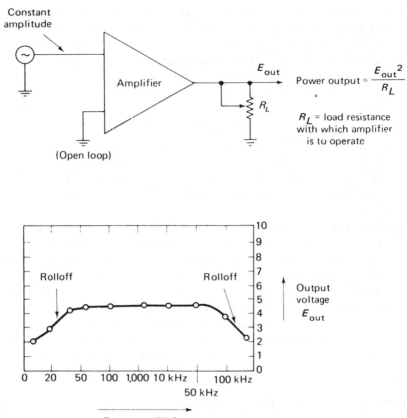

Fig. 8-12 Amplifier frequency-response test connections and typical response curve. (John D. Lenk, Handbook of Basic Electronic Trouble-shooting, © 1977, p. 172. Courtesy Prentice-Hall.)

input level is to monitor the amplifier output with the meter and increase the generator amplitude (set at a frequency of 1 kHz) until the amplifier is overdriven. This point is indicated when further increases in the generator amplitude do not cause further increases in meter reading. Set the generator output *just below* this point. Return the meter to monitor the generator amplitude (at the amplifier input) and measure the voltage. Keep the generator at this voltage *throughout* the test.

4. If the amplifier is provided with any operating or adjustment controls (volume, loudness, gain, treble, balance, bass, etc.), set these controls to some arbitrary point when making the initial frequency response measurement. The response measurements can then be repeated at different control settings if desired.

5. Record the amplifier output voltage on the graph. Without changing the generator output amplitude, increase the generator frequency by some fixed amount and record the new amplifier output voltage. The amount of frequency increase between each measurement is an arbitrary matter. Use an increase of 10 Hz where rolloff occurs and 100 Hz at the middle frequencies.

7. After the initial frequency response check, the effects of operating or adjustment controls should be checked. Volume, loudness, and gain controls should have the same effect all across the frequency range. Treble and bass controls may also have some effect at all frequencies. However, a treble control should have the greatest effect at the high end, whereas a bass control should have the greatest effect on the low end.

8. Note that the generator output amplitude may vary with changes in frequency, a fact often overlooked in making a frequency response test. Even precision laboratory generators can vary in amplitude with changes in frequency, thus resulting in considerable error. It is recommended that the generator amplitude be monitored after each change in frequency (some generators have a built-in output meter). Then, if necessary, the generator amplitude can be reset to the correct value. It is more important that the generator amplitude *remain constant* rather than being set at some specific value when making a frequency response check.

Amplifier Voltage Gain Measurement

Voltage gain in an audio amplifier is measured in the same way as frequency response. The ratio of output voltage to input voltage (at any given frequency or across the entire frequency range) is the voltage gain. Because the input voltage (generator output) is held constant for a frequency response test, a voltage-gain curve should be identical to a frequency response curve.

Power Output and Gain Measurement

The power output of an audio amplifier is found by noting the output voltage E_{out} across the load resistance R_L (Fig. 8-12) at any frequency or across the entire frequency range. Power output is $(E_{out})^2/R_L$.

To find the power gain of an amplifier, it is necessary to find both the input power and the output power. Input power is found in the same way as output power except that the input impedance must be known (or calculated). Calculating input impedance is not always practical in the case of some amplifiers, especially in designs where input impedance is dependent upon transistor gain. (The procedure for finding input impedance of an amplifier is described later in this section.) With input power known (or estimated), the power gain is the ratio of output power to input power.

In some applications, an *input sensitivity* specification is used. Input sensitivity specifications require a minimum power output with a given voltage input (such as 100 W output with 1 V RMS input).

Power Bandwidth Measurement

Many audio-amplifier design specifications include a power bandwidth factor. Such specifications require that the audio amplifier deliver a given power output across a given frequency range. For example, a circuit may produce full power output up to 20 kHz, even though the frequency response is flat up to 100 kHz. That is, voltage (without load) remains constant up to 100 kHz, whereas power output (across a normal load) remains constant up to 20 kHz.

Load Sensitivity Measurement

An audio-amplifier circuit of any design, especially power amplifiers, is sensitive to changes in load. An amplifier produces maximum power when the output impedance is the same as the load impedance.

The circuit for load sensitivity measurement is the same as the circuit for frequency response (Fig. 8-12) except that load resistance R_L is variable. (Never use a wire-wound load resistance. The reactance can result in considerable error.)

Measure the power output at various load-impedance versus output-impedance ratios. That is, set R_L to various resistance values, include a value equal to the amplifier output impedance, and then note the voltage and/or power gain at each setting. Then, repeat the test at various frequencies. Figure 8-13 shows a typical load-sensitivity response curve. Note that if the load is twice the output impedance (as indicated by a 2:0 ratio in Fig. 8-13), the output power is reduced to approximately 50 percent.

Dynamic Output Impedance Measurement

The load-sensitivity test can be reversed to find the dynamic output impedance of an amplifier circuit. The connections (Figs. 8-12) and the procedures are the same except that R_L is varied until *maximum* output is found. Power is removed, and R_L is disconnected from the circuit. The d-c resistance of R_L is equal to the dynamic output impedance. Of course, the value applies only at the frequency of measurement. The test can be repeated across the entire frequency range if desired.

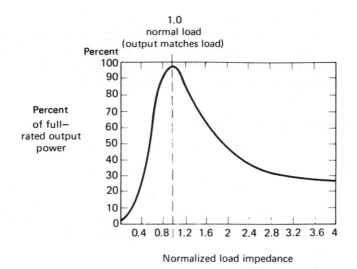

1.0
normal load
(output matches load)

Fig. 8-13   Output power versus load impedance (showing effect of match and mismatch between output and load). (John D. Lenk, Handbook of Basic Electronic Troubleshooting, © 1977, p. 175. Courtesy Prentice-Hall.)

Dynamic Input Impedance Measurement

To find the dynamic input impedance of an amplifier, use the circuit shown in Fig. 8-14. (Note that this same circuit and procedure can be used to find impedance of devices other than an amplifier. The procedure is an alternative to the impedance measuring technique described in Sec. 7-2 and shown in Fig. 7-3.) The test conditions for the circuit of Fig. 8-14 are identical to those for frequency response, power output, and so on. Move switch *S* between points *A* and *B* while adjusting resistance *R* until the voltage reading is the same in *both positions of S*. Disconnect *R* and measure the d-c resistance of *R*, which is then equal to the dynamic impedance of the amplifier input.

Accuracy of this impedance measurement is dependent upon the accuracy with which the d-c resistance is measured. A noninductive (not wirewound) resistance must be used. The impedance found by this method applies only to the frequency used during the test.

8-11   MEASURING AMPLIFIER NOISE AND HUM

If a meter is sufficiently sensitive, it can be used to measure the background noise level of an amplifier as well as to check for the presence of hum, oscillation, and so on. The meter should be capable of measuring 1 mV (or less),

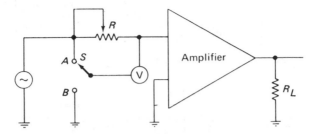

Fig. 8-14 Amplifier dynamic input-impedance test connection. (John D. Lenk, Handbook of Basic Electronic Troubleshooting, © 1977, p. 176. Courtesy Prentice-Hall.)

since this is the background noise level of some amplifiers. The basic procedure consists of measuring amplifier output with the gain control at maximum but without an input signal. Obviously, an oscilloscope is superior to a meter for noise-level measurement since the frequency and nature of noise (or other signal) will be displayed visually on the oscilloscope. The meter will show only the fact that a voltage is present and its amplitude. When measuring amplifier noise and hum with a meter, this procedure is applicable.

1. Connect the equipment as shown in Fig. 8-15. The load resistor R_L is used for power amplifiers and should have a value equal to the amplifier output impedance.

2. Set the meter to measure ac or "output." Use the lowest voltage scale.

3. Set the amplifier gain control to maximum and the tone controls to their normal position unless otherwise specified in the manufacturer's data.

4. Note the voltage level at the amplifier output. Temporarily disconnect the short at the amplifier input and note any change in output voltage. If the voltage level increases with short removed, the voltage is probably a result of pickup (external to the amplifier). If the voltage indication remains constant with or without the input short, the voltage is the result of background noise, oscillation, and so on (within the amplifier).

8-12 MEASURING FILTER-RESPONSE CURVES

A meter can be used to obtain a frequency-response curve of a filter. The basic method is the same as that for amplifier circuits described in Sec. 8-10. However, the test connections are usually somewhat different, and the response curves are quite different. Fig. 8-16 shows the test connections for measuring frequency response of a filter. Typical high-pass, low-pass,

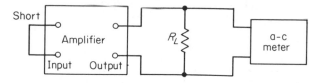

Fig. 8-15   Measuring amplifier noise and hum.

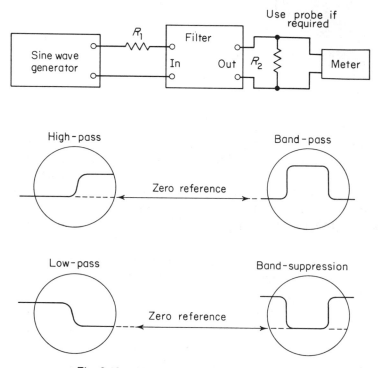

Fig. 8-16   Measuring filter response curves.

bandpass, and band-suppression response curves are also shown in Fig. 8-16.

Check the actual filter-response curve appearing on the graph paper (drawn as described in Sec. 8-10) against those of Fig. 8-16 or against the filter specifications.

Resistors R_1 and R_2 are included, since many test specifications for filters require that the input and output be terminated in their respective impedances. R_1 and R_2 may be omitted if not required by specification.

8-13 CHECKING OSCILLATOR CIRCUITS

The first step in checking any oscillator circuit is to measure both the amplitude and frequency of the output signal.

Oscillator Frequency Problems

The frequency is best measured using a frequency meter (counter) or possibly an oscilloscope. When you measure the oscillator signal, the frequency will be (1) right on, (2) slightly off, or (3) way off. If the frequency of a crystal-controlled oscillator is slightly off, it is possible to correct the problem with adjustment. Most crystal-controlled oscillators are adjustable; usually, the RF coil or transformer is slug-tuned, or there is a tuning capacitor. The most precise adjustment is obtained by monitoring the oscillator signal with a frequency counter and adjusting the circuit for exact frequency. However, it is also possible to adjust an oscillator using a meter. When the circuit is adjusted for *maximum signal amplitude*, the oscillator is at the crystal frequency. However, it is possible (but not likely) that the oscillator is being tuned to a harmonic (multiple or submultiple) of the crystal frequency. The frequency counter will show this, whereas the meter will not.

If oscillator frequency is way off, look for a defect rather than improper adjustment. For example, the coil or transformer may have shorted turns, the transistor or capacitor may be leaking badly, or the *wrong crystal* was installed in the right socket (this does happen).

Oscillator Signal Amplitude Problems

When you measure the oscillator signal, the amplitude will be (1) right on, (2) slightly low, or (3) very low. If the amplitude is slightly low, it is possible to correct the problem with adjustment. Monitor the signal with a meter (using the procedures of Sec. 8-8, or as described in the equipment service manual), and adjust the oscillator for maximum signal amplitude. This will also lock the oscillator on the correct frequency. If the amplitude is very low, look for defects such as low power-supply voltages, leaking transistor and/or capacitors, and shorted coil or transformer turns. Usually, when signal is very low, there will be other indications, such as abnormal voltage and resistance values.

Oscillator Bias Problems

One of the problems in troubleshooting or checking solid-state oscillator circuits is the bias arrangement. RF oscillators are generally reverse-biased, so that they conduct on half cycles. However, the transistor is initially forward-biased by d-c voltages. This turns the transistor on so that the collector circuit starts to conduct. Feedback occurs, and the transistor is driven into heavy conduction. During this time, a capacitor connected to the transistor base is charged in the forward-bias direction. When saturation is reached, there is no further feedback, and the capacitor discharges. This reverse-biases the transistor and maintains the reverse bias until the capacitor has discharged to a point where the fixed forward bias again causes conduction.

From a check or troubleshooting standpoint, the measured bias on a solid-state oscillator can provide a good clue to operation, *if you know how the oscillator is supposed to operate*, that is, if you know what the *average bias* is when the oscillator is operating properly. Sometimes, this information is available in the service literature. However, the one sure test of an oscillator is to measure output signal amplitude and frequency.

Oscillator Quick Check Using a Meter

It is possible to check whether an oscillator circuit is oscillating using a voltmeter and a large-value capacitor (typically 0.01 μ F or larger). Keep in mind that this procedure checks only that the circuit is oscillating; it does not prove that the oscillator is on-frequency and producing the correct amplitude signal.

Measure either the collector or emitter voltage with power applied to the oscillator (oscillator supposedly operating normally), and then connect the capacitor from base to ground as shown in Fig. 8-17. This should stop

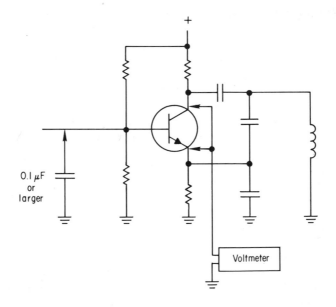

Fig. 8-17   Test connections for quick check of oscillator using a meter and capacitor. Voltmeter reading should change when capacitor is connected between base and ground.

oscillation, and the emitter or collector voltage will change. When the capacitor is removed, the voltage will return to normal. If there is no change when the capacitor is connected, the oscillator is probably not oscillating. In some oscillators, you will get better results by connecting the capacitor from the collector to ground. Also, do not expect the voltage to change on an element without a load. For example, if the collector is connected directly to B +, or if the emitter is connected directly to ground, these voltages will not change, with or without oscillation.

Index

Resistance of meter movements, 130
Resistive divider, precision, 31
Resistor tests, 157
Resolution, digital meter, 89
Resonant frequency of LC circuits, 195
Reverse-leakage tests, diode, 145
RF:
  amplifier alignment, 203
  probe, 14, 58
Ripple measurement, 176
RMS to dBM conversion, 83
RMS values, 26
RMS voltage, 14

Safety precautions, 72
Sawtooth waves, measuring, 141
Scales:
  AC, 79
  dB, 80
  DC, 77
  meter, 26
  ohmmeter, 75
Selectivity, receiver, 206
Self-resonance of coil, 198
Sensitivity, receiver, 208
Servicing with meters, 200
Shop meter, test and calibration, 118
Shunt:
  ammeter, 4
  fabricating, 133
  values, 132
  values, calculating, 130
Signal tracing, 59, 205
  in amplifiers, 212
Sinewave, 14
  measuring, 141
Solar battery test, 149
Special measurement procedures, 130
Squarewaves, measuring, 141
Staircase ramp digital meter, 38
Standard cell, 113
Student cell, 113
Suppressed-zero voltage measurements, 139
Swamping effect, 79

Test and calibration of shop type meters,
  118
Testing components, 144
Test leads, 87, 91
Testing meters, 111

Test prod, 55
Thermal resistor rests, 157
Thermocouple meter, 16
Three-phase measurement, 172
Tracing signals in receiver circuits, 205
Transformer:
  isolation, 72
  tests, 153
Transistor tests, 150, 193
Transistor voltages, 187
Transmitter circuits, servicing, 211
Troubleshooting capacitor circuits, 175
Troubleshooting with transistor voltages,
  190
True power, 29
Tunnel diode test, 147
Turnover effect, 80

UJT tests, 152

Variable resistor tests, 159
Varistor diode, 99
Voltage divider, 115
  probes, 57
  for probes, 67
Voltage (in circuit) measurement, 187
Voltage measurements, 87, 107
  digital meter, 93
  suppressed-zero, 139
Voltage sensitive circuits, 180
Voltage to frequency conversion, 40
Voltmeter, 5
  measurements, 107
  ranges, extending, 138
  test and calibration, 112
VOM, 2

Wattmeter, 29
Waves, complex, 139
Weston cell, 113
Wheatstone bridge, 10
Winding balance (transformer) tests, 157

Zener diode tests, 146
Zero adjustment, 91, 103
Zero center scale, 10, 79
  checking, 124